T0176451

The Multi-Governance of Water

SUNY series in Global Politics
James N. Rosenau, editor

The Multi-Governance of Water

Four Case Studies

Edited by

Matthias Finger,
Ludivine Tamiotti,
and
Jeremy Allouche

State University of New York Press

Published by
State University of New York Press

For information, address State University of New York Press,
194 Washington Avenue, Suite 305, Albany, NY 12210-2384

Production by Diane Ganeles
Marketing by Anne M. Valentine

Library of Congress Cataloging-in-Publication Data

The multi-governance of water : four case studies / edited by Mat[t]hias Finger, Ludivine
Tamiotti, and Jeremy Allouche.
    p. cm. — (SUNY series in global politics)
  Includes bibliographical references and index.
  ISBN 0-7914-6605-1 (hardcover : alk. paper) 0-7914-6606-X (pbk. : alk. paper)
  1. Water-supply—Economic aspects—Case studies. 2. Water-supply—Management—
International cooperation—Case studies. 3. Water resources development—International
cooperation—Case studies. 4. Watershed management—International cooperation—Case
studies. I. Finger, Matthias. II. Tamiotti, Ludivine (date) III. Allouche, Jeremy.
IV. Series.
  HD1691.M85 2005
  333.91'6217—dc22   2005001460

10 9 8 7 6 5 4 3 2 1

Contents

Illustrations

FIGURES

MAP

TABLES

Chapter 1

Introduction: Conceptual Elements

Matthias Finger, Ludivine Tamiotti, and Jeremy Allouche

Governance—as opposed to government—defines the phenomenon of societal problems (in our case water) appearing to be too interlinked, too complex, but also too overwhelming for any single nation-state to address them alone. Multi-level or simply multi-governance relates to the fact that such problems need to be tackled simultaneously at all relevant policy levels, i.e., from the local via the regional and the national to the supranational levels, and that these levels further need to be interconnected. For the purpose of defining a concept of multi-governance, case studies of the governance of four river basins have been conducted following a single conceptual framework. The basis of this conceptual framework is set in general terms in the introduction and then discussed and further developed in the conclusion that focuses on transboundary river basin governance.

The conceptual framework developed in this book builds on the analysis of the process of globalization, which has already—and independently of any particular issue—altered the way in which traditional politics works. This process has touched upon the way in which the State operates and is involved simultaneously both of the level above and below the nation-state, together with nongovernmental actors. The term *governance* therefore emerges as the concept that allows one to reconceptualize this changing role and functioning of politics. As such, "governance" defines a function—i.e., the function of collectively solving societal problems—, as opposed to government (local, national, and to a limited extent international), which defines a structure. In other words, this chapter, which is structured into four sections, aims precisely at doing this, i.e., defining how collective problem-solving is best conceptualized once the nation-state is considered to be too limited to warrant results.

1

In the first section of this chapter, we will present the implications of globalization on the nation-state. The argument, here, is that the nation-state is undergoing substantial changes as a result of globalization and that, therefore, governance emerges as a new phenomenon in order to solve collective problems, especially in the case of transboundary water issues and river basins. In the second section, we will discuss the two currently dominant theories in the area of governance of resources, namely regime theory and Common Property Resources Management Theory. The third section will then introduce the issue of water and river basins, as well as their management. The fourth section will summarize the overall conceptual framework for our study.

1. GLOBALIZATION AND THE IMPLICATIONS ON THE STATE

We would like to first position the concept of governance in general, and of multi-governance in particular, within the broader phenomenon of globalization. Indeed, irrespective of the issues at hand—which by themselves call for a governance approach, as we will see in the next section—globalization appears to have a significant influence on the nation-state and on its capacity to solve collective problems, even at the domestic level. In this section, we therefore want to show how traditional State authority, and especially problem-solving capacity, is being undermined by globalization, how governance in general, and multi-governance in particular, appear as a solution to this problem, and particularly so in the area of water management.

1.1. Globalization

The special importance of both the concept and the practice of governance stems from the fact that globalization has profoundly altered the premises of, and the ground rules for, traditional, i.e., nation-state-based, politics. As such, globalization has challenged the roles and functions of the traditional political actors (e.g., nation-states and related actors and institutions at local, national, and international levels) regulating public affairs and promoting industrial development. A situation has been created, where problems are no longer simply solvable at the nation-state level, where new often equally powerful actors have emerged parallel to the nation-state, and where new institutions both above and below the nation-state level are being created.

Quite logically, then, "governance" emerges as a new concept—and to a certain extent as a set of new practices—, seeking to capture this new politico-institutional reality. As a matter-of-fact, the concept of governance so far has mostly become prominent in the study of international public affairs.[1] There, the idea of governance seems to be in a good position to meet the multiple and interrelated challenges of the twenty-first century, as it indicates new, more cooperative ways of managing public affairs, i.e., solving collective problems (but still mainly among nation-states), capable of coping with the emergence of new actors and the evolution of traditional actors' roles, while capturing growing interdependencies.[2] While subscribing to the general line of this argument, this chapter will nevertheless examine the concept of governance critically, while seeing it in the context of changing politics at all levels and not just at the international level. In other words, the term *governance* will be used in a much more radical sense, seeing governance essentially as a new function of solving collective problems in a fragmented and multi-level political environment characterized, before all, by a multitude of actors and interests.

Indeed, the phenomenon of globalization—besides being characterized by growing interrelatedness, interdependency, complexity, and so forth—is also characterized by the slow but steady ascent of non-state actors. Among these, the two increasingly powerful ones are transnational corporations (TNCs) and nongovernmental organizations (NGOs): until recently, TNCs were considered to be the most typical of such new global (and local) actors.[3] Some of them—with their strategic vision, their mobility, and their economic and sometimes even political power— have already become more powerful than many governments.[4] But one often forgets that this same phenomenon of globalization can also be found in the not-for-profit sector.[5] Thus, one also has to increasingly deal with "multi-national" NGOs. In addition, new global agencies have emerged that are no longer entirely controlled by governments, as this is still the case of most United Nations agencies. We are thinking here in particular of the Bretton Woods institutions.[6] All of these actors not only have a global strategic vision, but are today among the most active promoters of globalization and of new governance mechanisms and arrangements.

These new actors span the entire gamut of societal levels, as they can be found simultaneously above and below the nation-state, and therefore also increasingly with the nation-state, as so-called partners. In other words, parallel to the emergence of new partners for the nation-state, new levels of collective-problem-solving below and above the nation-state are also emerging, often with corresponding institutions, within which the nation-state is no longer necessarily the dominant actor. Therefore, besides leading to the emergence of new non-state actors, globalization also leads to the emergence of new policy levels. In short, while globalization builds on previous historical trends of rationalization, institutionalization, expansion, and subsequent socio-cultural and ecological degradation and exploitation, it now seems to have reached a new stage in the form of new and institutionalized organizations, which increasingly span the entire range of levels, from the local via the national to the global. And this is precisely where the concept of multi-level or multi-governance comes in.

1.2. The Concept of Multi-(level) Governance

Having now located the concept of governance within the dynamics of globalization, we would like to precisely define, in this section, what we mean by *multi-level governance*. In doing so, we will first look more closely at the concept of governance, then define governance at the various levels, and finally link these levels in the concept of multi-level governance.

To recall, traditionally, i.e., since the French Revolution, the nation-state so-to-speak had had a monopoly over collective problems, even though it did not always manage to solve them. With globalization, however, this monopoly, along with many other State monopolies, erodes. As a result, it also becomes acceptable to define societal problems at levels other than the nation-state, especially at the global level, if one thinks, for example, of problems of peace, security, or environmental protection. Simultaneously, nation-states also increasingly push problem-solving downward to the local level, and often peoples themselves take the initiative to addressing societal problems locally even being explicitly encouraged by the State.

At the practical level, governance indeed refers to a mode of co-ordination of interdependent activities.[7] Governance thus can be understood as the establishment and operation of a set of rules of conduct that define practices, assign roles, and guide interaction so as to come to grips with collective problems.[8] Moreover, governance encompasses the various ways in which institutions, actors (public, private, and not-for-profit), resources, regulations, and mechanisms interact through a continuous process, in order to find cooperative solutions to vital societal functions. To quote Ernst-Otto Czempiel, governance means "the capacity to get things done without the legal competence to command that they be done."[9] This is what makes the difference between governance and government. Both of them are concerned with rules and collective action but with a difference in the processes used.[10] Governance is thus particularly appropriate to a situation where the nation-state loses its monopoly of legitimate power.

Governance therefore implies a stakeholder approach. Such an approach in public affairs probably became for the first time accepted at a global level within the context of the United Nations Conference on Environment and Development (UNCED). Ever since, mechanisms for stakeholder participation have become more and more prominent. For example, the 1992 United Nations Convention on Desertification has requested the private sector from underdeveloped countries to pay directly for national plans of action against desertification (Art. 6). Another example can be found in the process of the Inspection Panel of the World Bank. Under strict conditions, NGOs are allowed to file a suit for action before an organ composed of independent experts, the Inspection Panel, in order to assert their rights infringed by misconduct of the World Bank or of the Borrower State.[11] In other words, stakeholders are increasingly recognized ways of participating that go far beyond lobbying, which was the traditional approach to influencing state-centric politics.

There is a growing awareness that States are too large to solve some local and regional problems, and too small to address some global challenges.[12] In a sense, politics is becoming more polycentric with States merely one of the levels in a complex system of overlapping and often competing agencies of governance.[13] Such complexity implies the need for an integrated multi-level conceptualization of governance. However, so far we have primarily observed the emergence of governance practices, as well as corresponding conceptualizations, at the different levels taken separately. One increasingly sees governance-type of arrangements not only at global, but also at the (supranational) regional, and even at the local levels. Not to mention the fact that similar arrangements are now also emerging even at the nation-state level.[14] There does not appear to exist, so far, a coherent conceptualization of how these different levels of governance are being linked together, along the concept we would like to call "multi-level governance."

The level where governance has been most explicitly conceptualized so far is the global level, giving rise to the concept of "global governance." The argument for global governance is quite straightforward, as it stems from the observation of growing transnational operations and linkages and problems of global proportions, as well as from the inability or unwillingness of States to tackle these problems. So far, the major institutions tackling such global problems are still state-dominated institutions—for example, the UN and the Bretton Woods institutions. However, other non-state actors—such as transnational corporations and global non-governmental organizations—are also increasingly becoming recognized actors in this arena. Global governance is certainly the most prominent, but also the most vague use of the term *governance*. As a matter-of-fact, and since the early '90s, the notion of global governance became most popular in the field of international relations and institutional analysis.[15] Global governance is clearly rooted in the idea that economic and financial globalization have profoundly redistributed economic and political power, thus challenging State authority. Also, since the '90s, the development of humanitarian interventions has altered the previous basis of interstate order, allowing for suprastate actors to increasingly interfere into national and local matters. Such changes were conceptualized, albeit not systematically, by an international commission, which met on a regular basis in Geneva since 1992, the so-called Commission on Global Governance. In 1995, this commission published a report entitled *Our Global Neighborhood,* in which it submitted a wide range of proposals in various fields such as environmental governance, economic interdependence, or UN reform. In the framework of this report, the commission defined the concept of governance as follows:

> Governance is the sum of the many ways individuals and institutions, public and private, manage their common affairs. It is a continuing process through which conflicting or diverse interests may be accommodated and co-operative action may be taken. It includes formal institutions and regimes empowered to enforce compliance, as well as informal arrangements that people and institutions either have agreed to or perceive to be in their interest.[16]

With such a definition of governance, it is indeed possible to capture about everything, from individuals working together to cooperation among nation-states. Furthermore, it must be highlighted that this conceptualization of governance mixes together institutions and individuals, and does not account for their relative power, nor for their different strategic interests. As such, this is a conceptualization that is quite typical of UN jargon. Not astonishingly, it is also a particularly nonconflictual conceptualization of cooperation, inspired as it is by humanistic and New Age philosophies. The reality of "global governance" however looks significantly different, and makes such conceptualization look like wishful thinking. Some authors have furthermore tried to enlarge the concept of global governance, so that it can also include grassroots actors, as well as the role played by local peoples. In doing so, they elevate civil society actors to global players, thus not only confusing levels, but moreover ignoring the status and role of institutions. Again, such fuzzy thinking is quite typical of an intellectual tradition, which seeks to transform locally rooted peoples into a global civil society. One can see, says R. Lipschutz, the emergence of a multilevel and very diffuse system of governance, within which 'local' management, knowledge, and rule are of growing importance to coordination within domestic and international political 'hierarchies' as well as among regions and countries.[17]

This, however, is not to say that local levels and actions are not relevant when it comes to dealing with concrete issues and day-to-day concerns. Indeed, many international institutions and organizations now increasingly transfer their capacity to implement, as well as their ability to control the compliance, to local actors.[18] However, this seems to have less to do with an emerging global civil society, than with yet another instrumentalization of the local level by the global level, furthermore bypassing and weakening nation-states in the process. In no way can such civil society activities be conceptualized as an organized counterforce to newly emerging global actors.

It is worthwhile to mention, within this context of global governance, how the concept of governance is increasingly being used by some global actors, in particular by the World Bank and by some of the UN agencies. Indeed, the concept of "governance" has been used to a great extent by the *World Bank* since the early 1990s, and today it is rare to find a World Bank or a UN publication dealing with development issues, which does not refer to the concept of governance. However, the idea of governance in the framework of the World Bank entails a very specific content and definition. Indeed, World Bank working papers usually refer to the more eloquent concept of "good governance."[19] To recall, the concept of "good governance" has been introduced in order to address politically sensitive questions pertaining to State reform in developing and more recently in Eastern European countries. Such reform efforts, and thus the concept of good governance, most were of the time promoted by international financial institutions. Considering the fact that the statutes of organizations, such as the World Bank and the International Monetary Fund (IMF), expressly forbid them to take up political issues, the use of the concept of governance allowed these institutions to interfere into political and social questions without directly confronting the governments concerned, i.e., by defining governance in a quite technocratic way of (business) partners working together to promote investments and growth. Indeed, the World Bank has used the concept of good governance in a didactic manner, mainly in Africa, in order to designate the institutions and political practices that would be necessary for the (industrial) development of a given country.[20] Moreover, the concept of good governance has also been used, in the same perspective, within the context of the UNCED.

Here, the aim of the good governance approach is to create a political environment suitable for so-called sustainable development. As emphasized by K. Ginther, sustainable development requires a close interaction between government and peoples, and a lack of social structures entails crucial shortcomings of control and public accountability necessary in order to secure good governance.[21] More precisely, in the UNCED context, governance basically came to be defined as three things: (i) the participation of States in international law-making, (ii) the evolution of the decision-making mechanisms of international institutions, and (iii) the participation of nongovernmental entities in national and international decision-making and implementation processes.[22] Therefore, in both the World Bank and the UNCED contexts the notion of (good) governance appears to be very close to the notion of government, albeit a very technocratic form of government. Moreover, structural or good governance basically refers to broadly accepted structures of government, whose aim it is to promote the development of Western type "democracy." Governance then becomes a model able to provide nondemocratic or stateless countries with appropriate democratic institutions. Such governance does not address, for example, the interdependency and complexity of governance situations as we just defined them, i.e., in terms of collective problem-solving. Its only purpose is to define a certain way of operating State institutions, generally a way modeled after Western democracies, and aimed at developing an overall climate favorable for foreign direct investments.

The *regional* level—e.g., regional governance—is still primarily defined and articulated around States but the evolution of certain regional institutions indeed introduces a new level of governance, which is a supranational one. At the regional level, the main concern has been, so far, trade liberalization and economic integration. One can indeed witness the formal institutionalization of economic integration in various parts of the world through such institutions as the European Union (EU), the North American Free Trade Agreement (NAFTA), the Asia Pacific Economic Cooperation (APEC), the Association of Southeast Asian Nations (ASEAN), and the Southern Common Market (Mercosur). The European Union is in itself becoming a particular case in this regional governance process, due in particular to the depth of its economic integration, ultimately serving a political purpose. Indeed, the creation of a single market, and of a monetary union, makes the European Union a very different kind of supranational governance mechanism. Its concern to go further than just simply promoting economic integration with new emerging policies in the social, military, and foreign policy sphere shows how special this type of regional organization is as compared to the other cases. Besides these new regional economic type institutions, one can also find intergovernmental regional institutions comprising different UN specialized agencies and regional multilateral development banks (e.g., the African and Asian Development banks), which also play a substantive role in forging regional governance mechanisms. Nevertheless, one can foresee that most of these regional institutions focus mainly on economic concerns (i.e., development and trade and regional economic integration). We have to ask the question concerning the very nature and purpose of such regional governance mechanisms. Indeed, is regional governance another level of governance or simply an interstate strategic response to globalization? For example, some see in the European Union merely "an effort to contain the consequences of globalization."[23] As argued, for example, by Helen Wallace "European integration can be seen as a distinct Western European effort to contain the consequences of globalization. Rather than to be forced to choose between the national polity for developing countries and the relative anarchy of the globe, Western Europeans have invented a form of regional governance with policy-like features to extend the state and harden the boundary between themselves and the rest of the world."[24] In this sense, regional governance would basically mean reproducing the traditional nation-state.

It is of course at the *national* level where governments still have the strongest hold on solving collective problems. Nevertheless, even there, one can increasingly observe the erosion of traditional politics, both in terms of the policy process (policy-formulation, policy implementation, and compliance) and in terms of controlling operations. In the case of the policy process, it appears that more and more actors are being included in policy formulation, in the implementation, as well as in monitoring and compliance. Indeed, traditional "command and control" mechanisms are less and less effective. As a result, various stakeholders—e.g., businesses and NGOs—are being included both in the definition of the (environmental) policies and in their implementation. Indeed, national authorities are increasingly using private actors to implement and monitor their own national policies. For instance, in the field of wild bird protection, national authorities have established a strong partnership with scientific organizations to obtain accurate information on the implementation of their policies.[25] This evolution at the policy level is paralleled by a similar evolution at the operational level: indeed, and thanks to private sector participation and to other forms of outsourcing, many nongovernmental actors are now contributing to the implementation of public service objectives, thus necessitating all kinds of "governance-type" mechanisms in order to coordinate the various actors. This situation, however, is not (yet) comparable to the situation at the global level, where governments are clearly only one among many actors involved in managing public affairs. In this respect, the term *governance* at the national level is therefore not entirely appropriate. Indeed, governments retain their ultimate power, i.e., sovereignty over their territory, as well as control over legitimate power. Nevertheless, the capacity to get things done is increasingly dependent upon a government's ability to mobilize the various actors involved. Indeed, one must not forget that States are not unitary players: they have to deal with the pressure of the above and the below levels. The local groups can influence the definition of the strategy of a State at the international level. Some have characterized this phenomenon as a two-level game, in which the game played on the national level constrains the outcomes of the game played on the international level.[26] However, such local pressures do not necessarily weaken the state in the internal level-game. Governments can indeed use strong domestic oppositions to get better-off on the international level.[27]

At the *local* level, governance is again a totally different matter. In order to understand this type of governance, it is useful to recall that it is often at the local level where problems become first visible—even if they are only symptoms of global problems—, and where the resources and means to address these problems are very minimal. In other words, in the era of globalization, it is often at the local level where problems need to be solved as they arise. If these problems become too overwhelming, national, regional, and global actors also intervene. In noncrisis situations, however, national politics generally conceptualizes the local as an implementation problem: indeed, the local level is the final level of the implementation chain, i.e., the level at which all global, regional, and national policies will ultimately (have to) produce effects. This is also the level that ultimately provides the legitimacy for the entire public policy chain. Interestingly, and parallel to the erosion of traditional politics, the role of the local level is increasingly being recognized by both the national and global political actors, as being vital for their own effectiveness and legitimacy. To quote the Commission on Global Governance, some examples of governance at the local level may be a town council operating a waste recycling scheme; a multi-urban body developing an integrated transport plan together with user groups; or a local initiative of State agencies, industrial groups, and residents to control deforestation.[28] Many collective issues may indeed be handled more efficiently on the local level. Local populations often have intimate knowledge and experience of local ecosystems, as well as a sense of roots and continuity with a given place.[29] To recall, this level of governance has only been recognized in the context of larger global problems, i.e., in the late 1980s after the publication of the Brundtand Report. The need for, and the role of, local actions has moreover been enhanced by Agenda 21 agreed upon during the UNCED in Rio in 1992, which in turn encouraged local actors to develop their own local Agenda 21s. As a consequence, numerous initiatives all over the world were launched. Most of these initiatives are educational in nature and are not really problem-solving. In any case, such local actions are rarely self-contained local governance mechanisms, as they are part of a larger concept of implementing a "global public policy." However, what we would see as truly local governance is something else, namely community-based local problem-solving within the larger framework just outlined. Without doubt, such collective problem-solving efforts will become increasingly necessary parallel to the process of globalization and to the effects it has on local communities and on their livelihoods. It is therefore not surprising that it is at the local level where currently the most innovative governance practices and conceptual developments take place.

After having now outlined the emerging conceptualizations and practices of governance at the various levels, we must now turn to the concept of multi-level governance. In our view, *multi-level governance* simply refers to the fact that the emerging governance practices at the various levels—local, national, regional, and global—somehow need to be connected to one another. Such connection should furthermore occur in a more or less logical way, i.e., attributing the different levels of governance with the most appropriate functions. Furthermore, our idea of multi-governance also implies that there is some sort of mechanism to articulate, manage, and control the interlinkages between the different governance levels, an articulation that would have to be performed by an actor, which so far does not exist. In this book, we will illustrate this idea at the example of transboundary river basin management.

If one looks at the literature, one will tend to find the concept of multi-level governance exclusively used in the context of the European Union. Here, multi-level governance defines a conceptual framework whose function it is "to rectify the failure of previous theories to recognize the roles played by various actors on the European stage."[30] More generally, the concept of multilevel governance serves as a legitimation for an otherwise weak democratic justification of European institutions.[31] However, there are two basic ideas that can be taken from this quite Eurocentric conceptualization of governance, namely the idea of subsidiarity on the one hand, and the idea of a coherent articulation of the different levels on the other hand. Subsidiarity means that problems should always be solved at the lowest governance level and being shifted upward of this cannot be achieved. The idea of a coherent articulation of the different governance levels along subsidiarity principles furthermore suggest that there is an actor, in our case the European Commission, responsible for it. Nevertheless, we want to go beyond this European definition of the concept of multi-level governance, as we also want to include the global level in this overall governance approach. Furthermore, we think, as just argued, that the European Commission conceptualization of multi-level governance remains not only quite state-centric, but is furthermore ultimately top-down in nature. Indeed, we would want to be more open to the possibility that the coordination and articulation of these different levels of governance does not have to be "governed" by one single level, especially not by the regional one.

1.3. Water and the State

Nowhere in contemporary world politics is the need for effective governance more apparent than in the realm of global environmental issues. Sustainable environmental protection stands as the most challenging features for future generations. In this regard, governments are facing enormous pressures, from their citizens and from each other, to address problems of pollution, natural resource degradation, and ecosystem destruction. But in an ecologically interdependent world, acting unilaterally is often not an effective response to these problems. The result is that governments are forced toward collective action and cooperative behavior, in which they construct mechanisms for transnational environmental governance. Collective environmental management poses a severe challenge because it involves the creation of rules and institutions that embody notions of shared duties that impinge heavily upon the domestic structures and organization of states and that seek to embody some notion of a common good for the planet as a whole.

Governments, both individually and collectively, therefore face growing pressures to act cooperatively in response to environmental problems. However, one obvious set of pressures for global environmental governance comes from the poor fit between the world's political map and the current ecological problems. Rivers, watersheds, weather patterns, forests, deserts, and mountains rarely fit the logic of the territorially based nation-state system. As suggested by Jessica Tuchman Matthews,

> The majority of environmental problems demand regional
> solutions which encroach upon what we now think of as the

prerogatives of national governments. This is because the phenomena themselves are defined by the limits of watershed, ecosystem, or atmospheric transports, not by national borders. Indeed, the costs and benefits of alternative policies cannot be accurately judged without considering the region rather than the nation.[32]

For most authors, sovereignty inhibits environmental protection.[33] The 'sovereign-as-enemy' thesis is appealing on logical and historical grounds. First, the nation-state's territorial exclusivity approach compared to an integrated ecological approach appears to be mutually exclusive. Second, the modern State has been an agent or accomplice in ecological degradation across the globe. One such skeptical position was summarized by Richard Falk, writing at the time of the first wave of global environmental concerns in the early 1970s.

A world of sovereign states is unable to cope with engendered-planet problems. Each government is mainly concerned with the pursuit of national goals. These goals are defined in relation to economic growth, political stability, and international prestige. The political logic of nationalism generates a system of international relations that is dominated by conflict and cooperation. Such a system exhibits only a modest capacity for international co-operation and co-ordination. The distribution of power and authority, as well as the organization of human effort, is overwhelmingly guided by the selfish drives of nations.[34]

With respect to these institutional inadequacies, the State's role and predominance in dealing with transboundary water resources management has gradually been, at least on the theoretical level, put into question. Indeed, most of the literature on environmental governance considers the State an inadequate unit in managing natural resources and pollution, whether transboundary or not, and this is particularly true for water resources. The so-called world water crisis has also shown the necessity for new institutions and regulatory regimes. Indeed, no matter how hard the government of Bangladesh tries to prevent floods, different actions and policies in this respect cannot really work without the active cooperation of riparian states, in this case Nepal and India. Another example could be the protection of water resources from acid rain. No matter what the government of Canada does, it will be unable to protect its resources without the collaboration of the United States.

These interdependencies have led most water specialists to conclude that new governance mechanisms were needed for a sustainable approach to water resources management.

2. EXISTING THEORETICAL APPROACHES TO WATER GOVERNANCE

Such questioning of the role of the State have led to the development of new theoretical approaches to manage transboundary water resources. But these approaches are part of a much larger debate on the role and function of the nation-state in a globalized world. The main question of course is whether, in the era of globalization, the nation-state and corresponding politics, still is, and can be, a relevant actor of collective problem-solving.

Because of an increase in the number of international institutions, the growing interdependency and complexity from local to global issues, and the emergence of new actors on the international arena, it became necessary, during the 1990s, to redefine international public action. It is in this context that the concept of governance has been reintroduced, mainly in order to define *area-specific forms of governance*. Here, the term *governance* intends to analytically describe interdependencies and complexities involved in the operation of a given community or institution, generally limited to a geographic area and even to a specific issue (e.g., polar bears) or resource (e.g., water). The idea is not to focus anymore on the operation of structures, but rather to understand the forces and powers involved in the overall process "governing" an issue. In the context of international relations theory, the concept of area-specific governance has thus been used at two levels, one above and one below the State level, i.e., regime theory (above) and common property resources management theory (below). Both of these theories will now be discussed to show they have been applied to transboundary water governance.

2.1. Regime Theory

Regime theory became and still is an important tool in the international relations literature that helps explain the basic issues, challenges, and functioning of transboundary water governance. International relations specialists have focused their attention on regime theory, as well as on related subjects, such as institution effectiveness, implementation, and compliance mechanisms as analytic tools to explain the concrete functioning and complexity of the international system. *Regime* thinking has indeed initiated a new trend in the reflection on international institutions. During the 1980s, many studies have been carried out on international regimes, which are now fully part of the governance thinking. As Oran Young says, "international regimes, . . . are . . . specialized arrangements that pertain to well-defined activities, resources, or geographical areas and often involve only some subset of the members of the international society. Thus we speak of the international regimes for whaling, the conservation of polar bears. . . ."[35] The classic definition of an international regime is indeed that of a set of "principles, norms, rules and decision-making procedures around which actors' expectations converge in a given issue area."[36] Almost always, international regimes have at their core an international law agreement that established specific rules, commitments, and decision-making mechanisms to improve the process of governance.[37] Consequently, regimes are basically centered around nation-states, even though they can, and often do involve, other actors when it comes to regime formulation, implementation, and evaluation.

Furthermore, a number of scholars have focused their studies on the *effectiveness* of international regimes. The purpose of such an analysis is to enhance the appropriateness of rules to facts. Effectiveness can be defined as the degree to which international environmental agreements and organizations lead to changes in behavior that help solve collective problems.[38] However, one should notice that the effectiveness approach is a rather tricky one, as the assessment of the effectiveness of an international accord or regime is generally rather vague and difficult to establish. The effectiveness of regimes is here very similar to the effectiveness and corresponding evaluation of public policies at the national level.

In the same context, the study of *compliance* and *implementation* processes as part of international regimes are of growing concern to the operation of international institutions. Indeed, the nature and extent of international environmental commitments have been changing in recent years, since States take on more responsibilities under treaties and agreements that are increasingly stringent and with which they must comply.[39] Such a study on compliance and implementation issues is in direct correlation to the issue of effectiveness. Implementation and compliance are the processes by which the effectiveness of an accord, a policy, or a regime can be assessed.

Transboundary waters have often been cited as good illustrations of these new institutional arrangements. Supranational water commissions have actually been the first major form of international organizations and have therefore been the illustration of what a regime is.[40] Early models of freshwater regimes were mostly sectoral in nature. Most dealt only with individual issues that had emerged as a result of coordination needs among States sharing a river or a lake. The earliest agreement, for example, sought to establish rules for the freedom of navigation on international rivers. Progressively, matters such as apportionment of waters, flood control, irrigation, and energy generation took over as primary concerns in treaty making.

Nonetheless, the notion of water regimes has paradoxically been used in a very state-centric way, and this despite recent original studies highlighting the need for more advanced ecosystem regimes.[41] Most of these institutions, which are quoted as illustrations of these water regimes, are still very dependent upon States. Indeed, most of these regimes have no real discretionary power over transboundary waters. The International Commission for the Protection of the Rhine against Pollution, the International Joint Commission (United States and Canada), and the International Boundary and Water Commission (Mexico and the United States) are three typical examples. Each has the power to investigate pollution problems, for example, and to make recommendations but has no power to implement them. Despite the long history of the international commissions for shared freshwater resources, riparian states' sovereign logic is evident in the limits of these institutions. Among the numerous commissions in existence today, many continue to have a very narrow and technical focus or lack significant powers altogether. To date, no evolution toward institutional structures with significant original jurisdiction or powers for joint management of shared water resource has been implemented.

In short, regime theory pertains to a specific issue generally located at the supranational level, often involving the solution of a specific collective problem by means of nation-states' and of other actors' cooperation. To recall, such a regime is generally grounded in a legal framework, i.e., most often an international convention. It remains therefore somewhat state-centric. Yet, when it comes to transboundary water governance, there is clearly a need for a new approach that goes beyond the different aspirations and policies of the States, trying to regroup all of the major users in defining a more efficient and ecologically sustainable approach to transboundary water governance.

2.2. *Common Property Resources Management Theory*

A "common" is a resource to which no single decision-making unit holds exclusive title. This means that it is owned by no one (*res nullius*) or by everyone (*res communis*). Common pool resources are based on two attributes: the difficulty to exclude beneficiaries and the subtractibility of use. The crucial characteristics of common property resources are that property rights to parts of the resource cannot be defined and enforced. In this regard, some authors have considered water a common pool resource.

Common Property Resources Management Theory (CPRMT) looks at the effects of institutions on behaviors and outcomes in the area of (natural) resources. It is based on the resources to which a group of people have coequal rights, specifically rights that exclude the use of these resources by other people.[42] The purpose of CPRMT is to conceptualize how individuals and groups can organize themselves to govern and manage common property resources. Furthermore, it intends to develop a theory of self-organization and self-governance in a specific area, generally related to specific resources (e.g., forests, water, and fisheries).[43]

While CPRMT is generally limited to local levels and to specific areas, it nevertheless considers cases where individuals are dependent upon a given resource as a basis of their economic activity. In other words, every stakeholder is directly affected by what the others do, which means that each individual or group has to take into account the choices of the others when assessing personal choices.[44] CPRMT, as well as studies of community-based organizations, are highly relevant in order to understand this level and mechanism of (self-) governance. Nevertheless, CPRMT is basically a resources management approach to governance, conceptualized mostly at the local and at times at the sub-regional level. It operates with a stakeholder approach, yet makes very specific assumptions about the nature of the stakes, i.e., mainly economic ones. While theoretically and conceptually very solid, it is difficult to extrapolate CPRMT beyond the local level, beyond the area of natural resources, and beyond some basic assumptions on stakeholders' economic (and strategic) interests.

Applied to the water sector, a large body of work has focused on the factors that foster collective action in irrigation systems, as well as the conditions under which local institutions are employed to manage local water resources.[45] More particularly, the development of water users association are more and more viewed as a successful path in averting the so-called tragedy of the commons and in fostering sustainable development. For example, Vermillon lists nine property rights for water users associations, namely water rights, rights to develop crop and method of cultivation, rights to protect against land conservation, rights to infrastructure use, rights to mobilize and manage finances and other resources, rights of organizational self-determination, rights to select and supervise service provider, and rights to support services.[46] The decentralization and devolution of irrigation system management and these initiatives of collective action through water users associations regarding property rights certainly represent new approaches to sustainable local management and responsibility. Nevertheless, these institutional designs are clearly inapplicable at a higher level and actually are not intended to be so. River basin management is barely examined in CPRMT and those rare case studies only focus on domestic river basin management. But there again, these studies limit themselves to property rights and look at institutional designs on permit resource users. They are primarily concerned with the different possibilities for developing, modifying, contesting, or transferring water rights.[47]

Overall, the CPRMT literature does not therefore provide any useful mechanisms for transboundary river governance and ignores certain dynamics and interactions between the different levels of governance in these types of rivers. The favored approach of CPRMT around property rights does not offer any interesting perspective for transboundary river governance, since the property rights between the different governments do not constitute an issue at all. Moreover, the definition of property rights and rules over these types of rivers also clearly relies on the traditional authority of State legislation, although we have seen previously that transboundary water governance needs to go beyond the nation-state as the main unit in managing water resources.

In conclusion, the State in CPRMT is only indirectly present, yet it has a crucial function: indeed, it is the State—in particular its legal system—which defines the property rules, and, by doing so shapes the behavior of the different actors at the local level. It is also the State that ultimately enforces these rules. In short, in CPRMT the State is implicitly highly present, and this in a very traditional, i.e., legal fashion. This also means that CPRMT cannot be extrapolated beyond and above the nation-state level.

2.3. Conclusion

Both conceptualizations of governance we have mentioned here have their shortcomings. Common Property Resources Management Theory is conceptually very sound, but deals primarily with natural resources management at a subnational level and actually requires the State in its traditional functions in order to work. As such, it does not really address the issue of institutions and organized actors and their interests. Regime theory is, in our view, certainly the most interesting and promising approach to governance, as it identifies the supranational level and explicitly addresses the issue of organizations and institutions. However, regime theory refers to sectoral issues, and does not really constitute a comprehensive approach, influenced as it still is by the concern of how to implement public policies, albeit at an international level. Furthermore, regime theory still very much remains state-centric. Finally, both governance theories focus on one policy level without relating it to all other levels.

The multi-level approach is in fact much more suited to the current trend of globalization. The interdependencies between the different levels are of course of paramount importance in order to establish a collective problem-solving mechanism. Case studies will be designed from a river basin perspective since river basins transcends national boundaries, as well as provide for the link between the different conservation issues.

The management of international waters has long been an exclusive matter of bilateral relationships between States. The development of international organizations, the rise of global environmental issues, especially the ones related to river basin areas such as biodiversity, as well as the growing scarcity of waters, including desertification, have made the management of international river basins and water more generally an obvious global issue. Indeed, water management is related both to climate change and to biodiversity issues, not to mention economic development. If climate change has been introduced, so-to-speak, at a global level and is slowly moving down the "governance ladder," biodiversity, on the other hand, has mainly been introduced at a national level and is currently both moving upward and downward. Water, on the contrary, is clearly characterized by a bottom-up approach: awareness that it constitutes a governance issue has emerged first at a local level, tied as it is to livelihoods, and is currently moving to the national and to the global levels. We propose here to use the term *multi-governance* in order to account for the fact that not only does governance occur at all levels (from local to global), but that it also involves all stakeholders, and links several issues together with water such as biodiversity and climate change.

The reality, however, is that such a complete multi-governance approach (which will combine all actors and institutional levels) currently does not exist. Not surprisingly, the growing stress over water resources due to the intensive use of the shared waters for purposes of irrigation, livelihoods more generally, industrial production, electricity production, and many others more, raises the question of the current institutional structure for water governance. Many diverse actors are currently involved in managing, say, for example, a river basin, ranging from public authorities and international institutions to private corporations, public electricity boards, communities, and individuals. Consequently, a large number of rules apply and lots of conflicting interests interact at any given level of water management, i.e., from local to purely domestic to bilateral, multilateral, and even to global levels. The result of this proliferation of actors, actor-arrangements, rules, institutions, and even regimes carries a substantial risk of overlapping, contradiction, or simply confusion, leading to poor "water governance." In other words, there is clearly a need for some sort of an organized multi-governance approach to water.

3. WATER GOVERNANCE

Water has local, regional, national, and international characteristics. Water is local in its network, as each municipality around the word is, for example, responsible for supplying water to its inhabitants. It also is, at this level, an important vehicle for social and economic development. Water is regional and/or international in its natural setting, and as such cuts through different political frontiers. Water could also become international with its possible commodification. Finally, water is national in that the owner of the resources is generally the nation-state. This last characteristic is clearly artificial and has been introduced by human beings. The major governance challenge therefore lies in relating these natural characteristics with the humanly imposed one.

In this book, we will focus exclusively on the governance of transboundary rivers. Our definition of governance pertains to collective problem-solving and therefore there is a need to identify the different problems of transboundary water governance. Without anticipating the analysis of the different case studies, one could categorize transboundary water governance around three major and perhaps contrasting issues, i.e., an ecological, an economic, and a political one.

3.1. Transboundary Rivers: Facts and Figures

Before turning to these three issues, it is important to understand that the governance of transboundary water resources is a major problem. According to the register of international rivers of the now defunct UN Centre of Natural Resources, Energy and Transport, it was estimated in 1977 that there were 214 major shared international freshwater resources, 155 of which flow through two countries, and the rest through three or more countries. These figures certainly need to be updated especially since the breakup of the Union of Soviet Socialist Republics and of the Balkan States. For example, a recent World Bank report refers to "over 245 river basins," which serve about 40 percent of the world's population and half of its surface.[48] According to other estimations there are much more. A group of researchers has updated the UN register to list 261 international rivers, which cover almost half of the land surface of the globe.[49] L. Milich and R. G. Varady put the number of shared river basins at "more than 300."[50] The United Nations Commission on Sustainable Development, in its 1997 Comprehensive Assessment of the Freshwater Resources of the World, also estimated that the number of transboundary rivers has risen to over 300.[51] Despite these differences, transboundary water governance clearly emerges as an important international issue. This is all the more true since a recent study has shown that 70 percent of the world's land area can potentially be influenced by river basin development.[52] How the river basin is developed and managed will therefore have a major impact upon present and future living standards of its inhabitants, as well as on the basin ecosystem. Let us now discuss the main issues pertaining to the governance of transboundary waters.

3.2. The Ecological Approach to Transboundary Water Governance

To recall, water flows according to natural characteristics and does not respect administrative boundaries. Therefore, the question arises as to whether water should be managed and its management structures defined according to existing administrative boundaries or according to natural boundaries, usually taken to be river basins. Not astonishingly and from a pure water resources point, river basin management is therefore increasingly recognized as the most appropriate level for managing water resources. At the international level, the river basin management principle has its origins in the report of the International Law Association pertaining to the use of water of international rivers.[53] More recently, during the Rio Earth Summit, it was also argued that independent management of water by different water-using sectors is not appropriate, that the river basin must become the unit of analysis, that land and water need to be managed together, and that much greater attention needs to be paid to the environment.

A river basin is defined by its watershed area. At the highest elevation are the upper reaches where snowmelt or precipitation feed into narrow streams that rapidly descend a steep gradient. These upper reaches feed into a middle reach creating a 'mainstream' of the river. Floodplains, lakes, and swamps characteristically are found around the slow-flowing river mainstream. Below the mainstream is the lower reach, where the river meets the ocean. In the lower reach, saline and freshwater mix, silt settles, and a delta forms. Subsurface water flows including underground aquifers are also part of the basin. Therefore, integrated management of a river basin requires a thorough understanding of the physical, chemical, geologic, natural, and environmental resources of the basin.

This approach is increasingly put forward as the main model for managing water resources. Of course, river basin units have existed for a long time as shown in L. Teclaff's excellent monograph on the river basin in history and in law.[54] Nonetheless, there is currently a much more explicit policy in the different countries around the world in adopting river basins as the main ecological unit for managing water resources. In fact, the current approach goes even further, since it wishes to institutionalize river basin functions at this level. The European Union, for example, has recently adopted the Water Framework Directive, where the basin is definitely assumed as the basic territorial entity for elaborating "Management Plans." This ensures coherence with the aims of sustainability and conservation of aquatic ecosystems. In other regions of the world, this approach is also being followed and encouraged by development agencies and by international organizations. The so-called French model is usually put forward as the model to follow. Indeed, in order to obtain an integrated "river basin management," the French administration has designed six regional agencies, responsible for the general management of each river including investment, research and development, and information. There are in fact six river basin committees and six river basin financial agencies. The financial agencies are in charge of planning and macromanagement of the rivers. The river basin committees facilitate coordination among all of the different parties involved in managing water resources. They fall under the joint supervision of the Ministry of the Environment (for technical matters) and under the Ministry of Finance (exclusively for financial matters). This model has been put forward by the World Bank and also by specialized international water organizations, such as the World Water Council or the Global Water Partnership.

This change from a sectoral to a river basin management approach can only be understood in light of, as some would put it, the current "world water crisis." Indeed, the world is facing increasing problems of *water scarcity*: although 70% of the world's surface is covered with water, only 2.5% are in fact freshwater,[55] while development has led to a considerable increase in water use (nearly sevenfold increase in freshwater withdrawals since the beginning of the twentieth century).[56] The increasing demands are causing water stress in many areas of the world, even in some humid areas where rising demand or pollution have caused overutilization of the local resource.

The UN Committee on Natural Resources noted with alarm in 1997 that some 80 countries, constituting more than 40% of the world's population, were already suffering from serious water shortages.[57] According to the UN Commission on Sustainable Development, about

460 million people live in countries using so much of their water re-sources that they can be considered to be "highly water-stressed." A further one-quarter of the world's population live in countries where the use of water is so high that they are likely to soon move into situations of serious water stress.[58] In 1995, water availability was estimated to be 10,001 cu m per person per year, while as recently as 1970, it was 7,300 cu m.[59] Water scarcity is in fact a recent phenomenon. As put by the World Water Council, in the 1950s, only a handful of countries faced this problem. Nowadays, and according to the same organization, an estimated 26 countries with a population of more than 300 million suffer from water scarcity. Projections for the year 2050 show that 66 countries, comprising about two-thirds of the world's population, will face moderate to severe water scarcity.[60]

This ecological crisis has led decision-makers to search for a new approach to water resources management and thus the concept of river basin management has been put forward. As already discussed, this approach has been followed by the creation of new institutions, i.e.: river basin commissions, which play an increasingly important role in transboundary governance issues. But, along these new environmental preoccupations, transboundary water governance also includes impor-tant economic aspects as we will now see.

3.3. The Economic Approach to Transboundary Water Governance

Water is also an important element of economic and social develop-ment and this has been the case since ancient times. The earliest evi-dence for river engineering can be found in the irrigation canals of eight-thousand-year-old in Mesopotamia. However, it is really since the Industrial Revolution that one can see an exponential growth in water use. In terms of governance, it is important to identify the different uses of water for economic purposes and observe their current dynamic. The growing pressure on water resources and its increasing use for economic purposes indeed puts the issue of water sharing at the top of the transboundary governance issue.

The impressive increase in water demand is due to the combina-tion of different but interacting factors, such as the expansion and diversification of human utilization of water—for agricultural and in-dustrial, as well for domestic purposes—as well as the high rate of the world's population growth. As to the causes of the increase of human utilization of water, Prof. Herbert A. Smith, in his landmark monograph on international rivers published in the early 1930s, observed that "One of the most noteworthy features of the last hundred years has been the immense increase in the use of water. In part this has been due to changes in personal habits, and in part to scientific progress. These two causes continually interact."[61]

Suffice it to say that between 1900 and 1990 global withdrawals of water for human consumption increased by a factor of over six. Worldwide demand for freshwater grew from 579 km³ per year in 1900 to 4,130 km³ in 1990 according to UN statistics.[62] The main turning points are the years following the Second World War. As emphasized in the report on *World Water Resources and Their Use* prepared for the International Hydrological Program of the United Nations Educational, Scientific, Cultural Organization (UNESCO):

> The situation has drastically changed during the recent decades. In many regions and countries of the world, the unfavorable results of long-term, often unreasonable, man's activities, were discovered. This concerned the direct use of water resources and the surface transformations on river watersheds. To a large extent this was due to a drastic increase in global water withdrawal since the 1950s. In turn this increase was caused by the intensive development of production forces in all spheres of the world economy in the course of scientific and technological revolution. As compared with the previous decades, during 1951-1960, annual water withdrawal increased fourfold. This occurred because of the drastic expansion of irrigated areas, the growth of industrial and heat-power engineering water consumption, and the intensive construction of reservoirs on all the continents.[63]

Let us now see the main reasons for the growing importance of water in economic and social development during the past century.

The development of hydroelectricity has certainly brought an important change in the way in which water is managed. From then on, water is not only used for agriculture or domestic purposes, but also for industrial purposes. The first use of dams for hydropower generation was at the end of the nineteenth century. The World Commission on Dams estimated that by 1900, several hundred large dams had been built in different parts of the world, mostly for water supply and irrigation purposes.[64] And from 1940 on, multi-purpose projects began to spread in Asia, Australia, and South America. In Africa, this type of project did not become widespread until the 1960s. By 1949, about 5,000 large dams had been constructed worldwide, three-quarters of them in industrialized countries. By the end of the twentieth century, there were over 45,000 large dams in over 140 countries.[65] Figure 1 illustrates the different periods of dam building during the twentieth century.

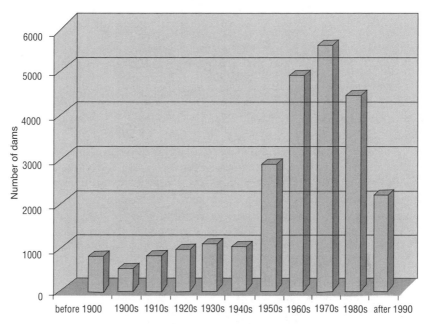

Source: ICLOD (1998) reproduced in World Commission on Dams (2002:9).
Information excludes dams in China

FIGURE 1. Construction of dams by decade (1900–2000)

One can see from this figure that the period of economic growth following the Second World War saw a phenomenal rise in global dam construction, lasting well into the 1970s and 1980s. The decline in the pace of dam building over the past three decades has been equally dramatic, especially in North America and Europe. However, some countries like China, India, Iran, and Turkey are still building dams.

There are also numerous other developments in other industrial sectors increasing the economic use of water. Today, industrial uses account for about 20% of global freshwater withdrawals. Of this, 57–69% is used for hydropower and nuclear power generation, 30–40% for industrial processes, and 0.5–3% for thermal power generation. The largest water withdrawals for manufacturing are made in the production of iron and steel, pulp and paper, petroleum products, chemicals, artificial silks and rayon, aluminum, explosives, and synthetic rubber. Like hydropower, one can see a slight decrease in industrial water use since the late '70s to early '80s. This can be explained by the fact that many countries have undertaken energetic measures to decrease industrial water withdrawal, and especially waste water discharge. These measures, reflect new environmental considerations, and in particular the wish to fight pollution.

Since the '70s to '80s, a tendency can be observed toward stabilizing and even decreasing industrial water withdrawal.

But the main development of the twentieth century regarding water has certainly been the so-called Green Revolution.[66] Of course, land irrigation has been practiced for millennia and the agricultural sector has always been the largest user of freshwater by far. However, industrial irrigation is mainly a phenomenon of the twentieth century. The Green Revolution is in fact characterized by a new movement to increase yields and to fight world hunger. The five major methods used were hybrid seeds, irrigation, fertilizers, pesticides, and mechanization. Research for the Green Revolution began in the 1940s when the Rockefeller Foundation launched a research project to improve agricultural yields in Mexico under the lead of the plant breeder Norman Borlaug. However, the Green Revolution rapidly became a world-wide agricultural movement. The high-yielding plants Borlaug introduced were a phenomenal success. Overall, there has been an increase in cultivated acreage of approximately 24% between 1950 and 1981.[67] Rice production techniques also changed and considerably increased, most notably in Asia.

The last major change in water use is the increase in domestic water consumption since the 1970s. Domestic water use is mainly re-lated to the quantity of water available for populations in cities. Indeed, urbanization is changing water consumption habits, as people are in-clined to consume more and more water in their daily lives. However, it should also be noted that domestic water use only makes up for a small share of water consumption, as compared to the agricultural and the industrial uses. Moreover, there are major differences between ur-ban people in developing and developed countries. It is indeed estimated that people in developed countries consume about ten times more water daily than those in developing countries. Having said this, the following figures on urbanization clearly reveal that water consumption and with-drawal will only increase in the near future. According to the report, *World Urbanization Prospects: The 2001 Revision* prepared by the United Nations Population Division, the world's urban population reached 2.9 billion in 2000, and is expected to rise to 5 billion by 2030. Whereas 30% of the world population lived in urban areas in 1950, the propor-tion of urban dwellers rose to 47% by 2000, and is projected to reach 60% by 2030. Table 1 summarizes this evolution toward urbanization.

TABLE 1. Total and urban population in the world (in billions)

	1950	1975	2000	2030
World population	2.52	4.07	6.06	8.27
Urban population	0.75	1.54	2.86	4.98

Source: Based on the United Nations Population Division (2001). World Urbanization Prospects: The 2001 Revision. New York: United Nations. Based on Table 2.6.

In short, from the end of the nineteenth century until the end of the twentieth century, water use considerably increased. Figure 2 and Map 1 show the different categories of water use from a chronological perspective since the beginning of the twentieth century.

The major conclusion is that the economic use of water has escalated since the beginning of the twentieth century with the development of hydroelectricity and irrigated agriculture. Water has therefore become more and more a vehicle for the development of local and national economies. However, these economic developments have given rise to a greater competition between the different users. Water is used simultaneously for navigation, irrigation, electric power, and for the supply of large cities, and the relative priority of these claims raises problems of great difficulty and importance. An economic approach by means of pricing is more and more put forward as the solution to these competing claims.

Moreover, these economic developments have important consequences on the quality of water. Some of this diverted water cannot replenish the watercourse. Other uses such as waste discharge or industrial sector use (particularly in the chemical industry) can indeed alter the quality of water. One can therefore see that these developments confront the governments with important responsibilities with regards to ecological and economic governance and especially in the international arena.

3.4. The International Relations Approach to Water Governance

Transboundary water governance also has an important political component. This is all the more true since international customary law over transboundary rivers is far from being clear. The recourse to international legal principles has been used in order to try to limit conflict over water resources. Nevertheless, as noted by Peter Gleick,

> Almost all of these [international law] focus on attempting to limit the environmental impacts of conflicts and war; few efforts have been made to address the equally important problem of limiting the use of the environment as an instrument of conflict, preventing conflicts over access to resources, or averting military responses to the consequences of environmental damages, such as population displacements.[68]

No international agreement ratified by a sufficient number of countries exists today to indicate which legal rule governs these rivers. And this still open question has evolved into different legal doctrines, as shown in Table 2.

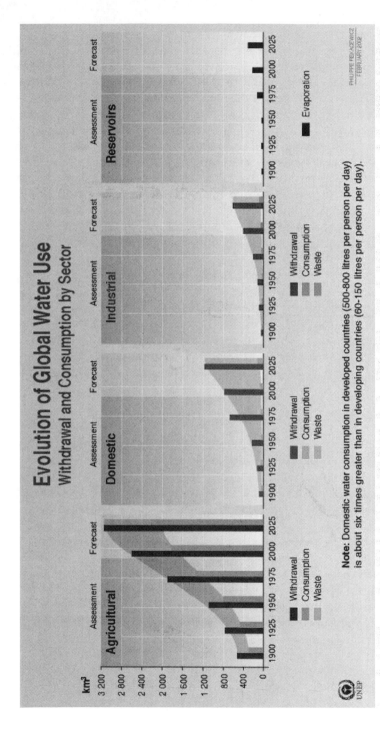

FIGURE 2. Evolution of global water use

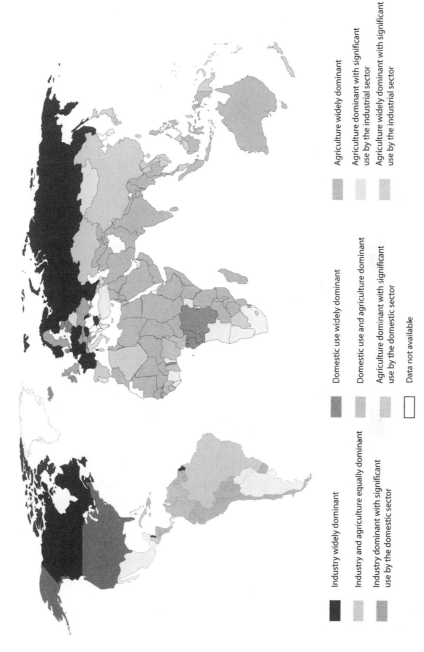

Industry widely dominant

Industry and agriculture equally dominant

Industry dominant with significant use by the domestic sector

Domestic use widely dominant

Domestic use and agriculture dominant

Agriculture dominant with significant use by the domestic sector

Data not available

Agriculture widely dominant

Agriculture dominant with significant use by the industrial sector

Agriculture widely dominant with significant use by the industrial sector

Map 1. Global freshwater withdrawal

TABLE 2. Conflict management doctrines on international rivers

---

(a) Absolute Territorial Sovereignty: Harmon doctrine

According to the reasoning behind this doctrine,[69] a State may adopt all of the measures deemed suitable to its national interest in regards to watercourses within its territory, irrespective of their effects beyond its borders. Accordingly, it may freely dispose of waters flowing within its territory, but cannot demand the continued free and uninterrupted flow of water from upper-basin States.

Proponents of this doctrine argue that an international water course within the territory of a State constitutes an integrated part of the public domain of that State. And since a State has full control over its own territory, other States acquire rights only with the agreement of the first State. This doctrine clearly favors upper-basin States.

(b) Absolute Territorial Integrity (or riparian rights theory)

This model is the direct opposite of the theory of absolute territorial sovereignty. It says that lower riparian States have the right to continued and uninterrupted (or natural) flow of the water from the territory of the upper riparian (basin) State. The theory is thus favorable to the lower-basin State.

The theory is sometimes criticized because it allocates rights without imposing corresponding duties. It has been invoked in situations where the continued flow of waters was critical to the survival of the state concerned, as in, for example, the case of Iraq and the river Euphrates.

(c) Limited Territorial Sovereignty and Limited Territorial Integrity

Theories of limited territorial sovereignty and limited territorial integrity are in practice complementary, even identical. Both say that every State is free to use the waters flowing into its territory, on the condition that such utilization does not prejudice the territory or interests of other States. In short, they say that States have reciprocal rights and obligations in the utilization of the waters of their international drainage basins.

(d) Community of interests in the waters

Some authors argue for a 'community' approach (i.e., State boundaries should be ignored and a drainage basin be regarded as an economic and physical unit). No State, for example, should dispose of the waters without consultation or cooperation with the other States (as is the case of many rivers, e.g. the Senegal River). The doctrine claims that the water system ought to be managed as an integrated whole. Such considerations lead to the implementation of basin-wide development programmes, jointly designed by all the riparian States in the river basin.

(e) The doctrine of Equitable Utilization

This model has evolved gradually as a result of the conflict among the competing theories just discussed (a, b, c, and d). It proposes that each basin State has a right to utilize the waters of the basin and as such is entitled to a reasonable and equitable share of the basin water.

TABLE 2. Conflict management doctrines on international rivers *(continued)*

The principle of equitable utilization reflects three fundamental concerns. First, it takes into account the socioeconomic needs of the basin States through an objective consideration of various factors and conflicting elements relevant to the use of the waters. Second, it aims at distributing the waters among the basin States in such a manner as to satisfy their needs to the greatest possible extent. And third, it seeks to distribute the waters so as to achieve the maximum benefit for each co-basin State with the minimum detriment.

Today, the model of 'limited territorial sovereignty' is probably the prevailing one in international watercourse rights and obligation since it does not downplay the interests of neither upstream nor downstream users, and recognizes at the same time the rights and obligation toward other riparians.[70] Moreover, the adoption of the UN Convention on the Law of the Non-Navigational Uses of International Watercourses in 1997 indicates that State practice regarding the utilization of transboundary water for nonnavigational use will follow the doctrine of equitable utilization.[71] However, as we will see in our different case studies, although many governments recognize the usefulness of this last concept, State practice over transboundary water resources seems to be governed by other imperatives, especially economic ones, as underlined in the previous paragraphs, but also by political ones. Indeed, water is more and more used as an instrument of foreign policy relations. Also, the importance of the environmental factor in international politics is more and more recognized. In 1965, a study by Harold and Margaret Sprout identified the environment as one factor influencing a nation's foreign policy.[72] Concerning transboundary watercourses, each State has in fact a dual preoccupation. On the one hand, it wishes to control its own water resources flowing through its country. It also wants to avoid the situation whereby the upstream users imperil its current and future social and economic development. In this regard, each State is ambitious to control and to use most of the transboundary water resources available. On the other hand, the State is also preoccupied with the quality of water and monitors whether the actions and industrial and agricultural developments in the upstream States do indeed affect the quality of its own waters. Both aspects are important in the political relations between the riparian states and can be understood along an upstream/downstream axis. Transboundary watercourses are in fact nearly always characterized by an unequal power relationship between the upstream and the downstream country. River basins have the interesting property that both positive and negative externalities usually have their effects only in one direction, that is, downstream. An upstream country affects the volume or quality of a downstream country's water by diverting or polluting it, but the downstream country cannot do the reverse.

In the situation of war, water can also become instrumental. Even though water was more of a war tool in ancient times as compared to nowadays, one can see that dams are increasingly becoming easy and primary targets of strategic military thinking. During the past century, hydroelectric dams were bombed during World War II, and the centralized dams on the Yalu River serving North Korea and China were attacked during the Korean War. Other examples can be found during the Iran-Iraq War, the 1991 Persian Gulf War, and the Yugoslavian Civil War.[78] During the Bosnian War, the Serbs immediately shut off Sarajevo's electricity and with it the city's water pumps. In order to get water, residents then had to line up at wells around the city, making them easy targets for Serb snipers and mortar shells.[74] The ability of Turkey to shut off the flow of the Euphrates, even temporarily, was a constant political and military stress at the beginning of the Persian Gulf conflict. While no such action was ever taken, the threat of the 'water weapon' was clear.[75]

3.5. Conclusion

Transboundary water governance is therefore characterized by three main and often contradicting approaches, i.e., an ecological, an economic, and a political one. These three approaches have an impact on transboundary water governance but are happening at different levels. For example, the ecological approach is clearly concerned with the river basin level. On the contrary, the economic approach is developed at the local, regional, and national levels. Finally, the political approach is concerned with the river basin. All of these approaches of course need to be coordinated in order to foster a better management of the basin. This book argues that multi-level governance is the most appropriate way of managing transboundary water resources.

4. FRAMEWORK OF OUR STUDY

The following conceptual framework was proposed to the authors of the case studies as a guiding thread. In order to address every aspect of the multi-governance of water in each river basin studied, three dimensions had been proposed: an understanding of the general context of the case, an analysis of institutions and actors relevant to the governance of the basin, and finally an explanation of the actual management of the river basin considered. In other words, the basin and not the boundary was taken as the relevant unit of analysis.

Concerning the first dimension, i.e., the general context of the water basin, the idea was to highlight every issue of the river basin, and to see which reflects the need for a multi-governance approach. This dimension thus accounts for the biological and physical characteristics of the river basin. For example, such facts as the presence of hydroelectric power generation, the navigation uses, the agricultural and urban needs, or the waste disposal mechanisms are of high relevance to the governance of the river basin. This general context also introduces the history of the river basin in terms of cooperation and conflicts among States, of decolonization, and of ecological and economic issues. Finally, multi-governance is introduced when it comes to considering the stakes of management of the river basin. These are typically manifold, i.e., geopolitical, environmental, and economic.

The second dimension addressed by the authors of the case studies is the analysis of the institutions and actors relevant to the process of multi-governance. It focuses on the identification of every institution (in its broad sense) and those actors that are involved in the management of the river basin, and this on two different levels: first horizontally, by pointing out the different institutions and actors involved at the local, regional, national, and international levels; and then vertically, by highlighting the relationships and the interactions, formal as well as informal, which exists between all of these levels. This institutional and actor dimension thus pertains to the different types of actors and institutions involved in water governance at one or several of these levels, i.e., political actors, public administrations, businesses, NGOs, and civil society actors. Many of these have specific sectoral orientations, referring to either the environmental, economic, cultural, religious, military, or scientific sectors. The process of identifying the institutions and actors involved has been envisaged in a systemic approach, i.e., by taking into account the simultaneous role of the different UN and other multilateral international agencies, the regional organizations, the secretariats of the international and regional agreements, the nongovernmental organizations, and the multi-national firms. The study of the relationships and interconnections between institutions is typically where the concept of multi-governance can be applied to address the complexity of a comprehensive management of the river basin. Such situations can then be qualified in terms of overlap, cooperation, partnership, or competition between agencies.

Finally, the third dimension useful to the understanding of the multi-governance of a river basin is the actual assessment of the management of the concerned river basin. The authors placed the current situation into a historical perspective, by considering the past, the present, as well as the general progress that had been realized so far.

From these three dimensions, we will propose, in the conclusion, four analyses assessing the concept of multi-governance for water management in the cases of the Danube, the Euphrates, the Mekong, and the Aral Sea. For each case, we will try to identify a specific model of governance that will then serve as a basis for our conceptualization of multi-governance. In other words, the main aim of this book is to create a conceptual foundation for approaching "multi-level governance of water" within a context of globalization. As such, it has to allow for two things (and should be judged accordingly), namely (i) to position water within a larger framework of governance and globalization, and (ii) to help conceptualize multi-governance as a means to address issues of water management at all interlinked levels.

NOTES

1. See Streck, C. (2001). "Global Public Policy Network, International Organizations and International Governance." Heinrich Boell Foundation. Washington DC (November), as well as other publications available from the Global Public Policy Institutions team.

2. Young, O. (1994). *International Governance. Protecting the Environment in a Stateless Society*. Ithaca NY: Cornell University Press: 4.

3. For instance, Barnet, R. & J. Cavanagh, (1994). *Global Dreams. Imperial Corporations and the New World Order*. New York: Simon & Schuster; Korten, D. (1995). *When Corporations Rule the World*. West Hartford CT: Kumarian Press.

4. Anderson, S. & Cavanagh, J. (2000). *The Rise of Corporate Global Power*. Washington DC: Institute for Policy Studies.

5. Lipschutz, R. (1996). *Global Civil Society And Global Environmental Governance*. Albany: State University of New York Press Press; Princen, T. & Finger, M. (eds.) (1994). *Environmental NGOs in World Politics: Linking the Local to the Global*. London: Routledge.

6. For instance, Chossoudovsky, M. (1997). *The Globalisation of Poverty. Impacts of IMF and World Bank Reforms*. London: Zed Books.

7. Jessop, B. (1998). "The Rise of Governance and the Risks of Failure: The Case of Economic Development." *International Social Science Journal* 155: 29 (March).

8. Young (1994). *op. cit.*: 3, 15.

9. Czempiel, E.-O. (1992). "Governance and Democratization." In Rosenau, J.N., Czempiel, E.-O. (eds.). *Governance without Government: Order and Change in World Politics*. Cambridge: Cambridge University Press: 250.

10. Stoker, G. (1998). "Governance as Theory: Five Propositions." *International Social Science Journal* 155: 17 (March).

11. Shihata, I. (1994). *The World Bank Inspection Panel*. Published for the World Bank. Oxford: Oxford University Press: 127.

12. Mische, P. (1992). "National Sovereignty and Environmental Law." In Bilderbeek, S. (ed.). *Biodiversity and International Law. The Effectiveness of International Environmental Law.* Amsterdam: IOS Press. 105–114: 110.

13. Hirst, P. & Grahame, T. (1996). *Globalization in Question.* Cambridge: Cambridge University Press: 183.

14. See, for example, Rosenau, J. (1995). "Governance in the Twenty-First Century." *Global Governance* 1: 13–43: 16–19.

15. For example, Paolini, A. J., Jarvis, A., & Reus-Smith, C. (eds.) (1998). *Between Sovereignty and Global Governance. The United Nations, the State and Civil Society.* New York: St. Martin's.

16. Commission on Global Governance (1995). *Our Global Neighbourhood.* Oxford: Oxford University Press: 2.

17. Lipschutz (1997). "From Place to Planet: Local Knowledge and Global Environmental Governance." *Global Governance* 3: 83.

18. Tamiotti, L. & Finger, M. (2001). "Environmental Organizations: Changing Roles and Functions in Global Politics," *Global Environmental Politics* 1.

19. World Bank (1992). *Governance and Development.* Washington DC: World Bank.

20. Senarclens, P. (1998). "Governance and the Crisis in the International Mechanisms of Regulation." *International Social Science Journal* 155: 92.

21. Ginther, K. (1995). "Sustainable Development and Good Governance: Development and Evolution of Constitutional Orders." In Ginther, K., Denters, E., & de Waart, P. (eds.). *Sustainable Development and Good Governance.* Dordrecht, Netherlands: Martinus Nijhoff: 155.

22. Sands, P. (1994). "International Law in the Field of Sustainable Development." *British Yearbook of International Law*: 355–356.

23. Wallace, H. (1996). "Politics and Policy in the EU: The Challenge of Governance." In Wallace, H. & Wallace, W. (eds.). *Policy-Making in the European Union.* Oxford: Oxford University Press: 17.

24. Ibid.,16.

25. Tamiotti & Finger (2001). *op cit.*: 1.

26. Putman, R. D. (1998). "Diplomacy and Domestic Politics: The Logic of Two-Level Games." *International Organization* 42: 427.

27. Ibid.: 440.

28. Commission on Global Governance (1995). *op. cit.*: 2.

29. Mische (1992). "National Sovereignty and Environmental Law.": 111.

30. Sloat, A. (2001). "Multi-Level Governance: An Actor-Centered Approach." Multi-Level Governance Conference. United Kingdom: University of Sheffield. June: 1.

31. Harnisch, A. (2002). "Multi-Level Governance beyond the Nation-State: The End of Legitimate Democratic Politics?" *Bologna Center Journal of International Affairs* (Spring) (available online).

32. Tuchmann M. J. (1989). "Redefining Security." *Foreign Affairs* 68 (2): 162–177: 166 (Spring).

33. On this subject, see Frank, D. J., Hironaka, A. & Schofer, E. (2000). "The Nation-State and the Environment, 1900–95." *American Sociological Review* 65 (1): 96–116; Hurell, A. (1994). "A Crisis of Ecological Viability? Global

Environmental Change and the Nation-State." *Political Studies* 42: 146–165; Soros, M. (1986). *Beyond Sovereignty: The Challenge of Global Policy.* Columbia: University of South Carolina Press; Hough, R. L. (2003). *The Nation-State: Challenges and Prospects.* Washington: University Press of America: chap. 5.

34. Falk, R. (1971). *This Endangered Planet: Prospects and Proposals for Human Survival.* New York: Vintage Books: 37–38. On the same view, see Dryzek, J. (1987). *Rational Ecology.* Oxford: Blackwell: chap. 6.

35. Young (1989). *International Cooperation: Building Regimes for Natural Resources and the Environment.* Ithaca, NY: Cornell University Press: 13.

36. Krasner, S. (ed.) (1983). *International Regimes.* London: Cornell University Press.

37. Victor, D., Raustiala, K., & Skolnikoff, E. (eds.) (1998). *The Implementation and Effectiveness of International Environmental Commitments. Theory and Practice.* Cambridge (Massachusetts) and London (England): IIASA/MIT Press: 8.

38. Ibid., 1.

39. Sands, P. (1996). "Compliance with International Environmental Obligations: Existing International Legal Arrangements." In Cameron, J., Werksman, J., Roderick, P. (eds.). *Improving Compliance with International Environmental Law.* London: Earthscan: 49.

40. Otto Popper talked about the international regime of the Danube. Popper, Otto (1943). "The International Regime of the Danube." *Geographical Journal* 102 (5/6): 240–253 (November–December).

41. Brunnee, J. & Toope, S. J. (1997). "Environmental Security and Freshwater Resources: Ecosystem Regime Building." *American Journal of International Law* 91 (1): 26–59 (January).

42. World Bank (1992). *op cit.*: 2.

43. Ostrom, E. (1990). *Governing the Commons. The Evolution of Institutions for Collective Action.* Cambridge: Cambridge University Press: 27.

44. Ibid., 38.

45. See, for example, Samakande, I., Senzanje, A., M. Mjmba Samakanda, I., Senzanje, A. & Mjmba, M. (2002). "Smallholder Irrigation Schemes: A Common Property Resource with Management Challenges." Presented at The Commons in an Age of Globalisation. The 9th conference of the International Association for the Study of Common Property. Victoria Falls. Zimbabwe, June 17–21.

46. Vermillon, D. L. (1999). "Property Rights and Collective Action in the Devolution of Irrigation System Management," www.capri.cgiar.org/status.asp.

47. See, for example, Allen, B. & Edella, S. (2000). "Convenant Institutions and the Commons: Colorado Water Resource Management." International Association for the Study of Common Property. Bloomington IN, May 31–June 4.

48. Salman, S. M. A. & Chazournes, L. B. (eds.) (1998). "International Watercourses—Enhancing Cooperation and Managing Conflict." *World Bank Technical Paper* 414: vii.

49. Wolf, A. T., Natharius, J. A., Danielson, J. J., Ward, B. S., & Pender, J. K. (1999). "International River Basins of the World." *International Journal of Water Resources Development* 15 (4): 387–427.

50. Milich, L. & Varady, R. G. (1999). "Openness, Sustainability, and Public Participation: New Designs for Transboundary River Basin Institutions," *Journal of Environment and Development* 8 (3): 258–306: 259.

51. United Nations, Commission on Sustainable Development (1997). *Comprehensive Assessment of the Freshwater Resources of the World*. Report of the Secretary General. U.N. Doc. E/CN.17/1997/9. New York: United Nations: paragraph 115.

52. Scudder, T. (1994). "Recent Experiences with River Basin Development on the Tropics." *Natural Resources Forum* 18 (2): 101–114.

53. Report adopted in the 48th conference of the International Law Association. New York, 1958.

54. Teclaff, L. (1967). *The River Basin in History and Law*. The Hague: Martinus & Nijhoff.

55. United Nations. Commission on Sustainable Development (1997). *op. cit.*: paragraph 33.

56. Gleick, P. H. (1998). *The World's Water: The Biennial Report on Freshwater Resources, 1998–1999*. Washington DC, Covelo, CA: Island Press.

57. United Nations. Commission on Sustainable Development (1997). *op. cit.*: paragraph 43.

58. Ibid.: paragraph 43.

59. Ibid.: paragraph 34.

60. World Water Council (2001). "World Water Challenges for the Twenty-first Century." http://www.worldwaterforum.org/Pressreleases/press1.html.

61. Smith, H. A. (1931). *The Economic Uses of International Rivers*. London: King & Son: 1.

62. Economist (1995). "Water: Flowing Uphill." August 12, 36, graph, citing the Food and Agriculture Organization of the United Nations.

63. Shiklomanov, I. (1999). *World Water Resources and their Use*. Joint SHI/UNESCO Product: Introduction.

64. World Commission on Dams (2002). *The Report of the World Commission on Dams*. London: Earthscan, chap. 1: Water, Development, and Large Dams: 8.

65. Ibid.

66. Term first used by the U.S. Agency for International Development Director William Gaud in March 1968. On the Green Revolution, see Germain, R. (1979). "La révolution verte: Ses origines, ses succès, ses contraintes." *Bulletin des Scéances de l'Académie Royale des Sciences d'Outre-Mer* 25 (4): 649–662; Thompson, K. W. (1973). "The Green Revolution." *Center Magazine* 6 (6): 56–66. On criticism on the Green Revolution, see Glaeser, B. (1987). *The Green Revolution Revisited: Critique and Alternatives*. Winchester MA; Shiva, V. (1991). *The Violence of the Green Revolution: Third World Agriculture, Ecology and Politics*. London: Zed Books.

67. World Watch Institute (1997). "Vital Signs 1997." Washington DC: World Watch Institute.

68. Gleick, P. (1993). "Water and Conflict: Fresh Water Resources and International Security." *International Security* 18 (1): 79–112: 105.

38   FINGER, TAMIOTTI, AND ALLOUCHE

69. The Harmon doctrine was named in honor of Attorney General Judge Harmon who first opined the idea in 1895 during an international legal dispute with Mexico over the Rio Grande.

70. Caflsich, L. (1989). "Règles générales du droit des cours d'eau internationaux." *Recueil des cours de l'Académie de droit international de La Haye* 219 (7): 9–225: 55, McCaffrey, S. C. (2001). *The Law of International Watercourses: Non-Navigational Uses.* Oxford/New York: Oxford University Press. Oxford Monographs in International Law: 137.

71. Table 2 has put the doctrine of equitable utilization in opposition with the doctrine of limited territorial sovereignty. For some authors like Caflisch, the doctrine of equitable utilization is the practical manifestation of the doctrine of limited territorial sovereignty. (Caflsich [1989]. *op. cit.*: 164).

72. Sprout, H. & Sprout, M. (1965). *The Ecological Perspective on Human Affairs with Special Reference to International Politics.* Princeton: Princeton University Press.

73. See Gleick (1998). *op.cit.*: 87–88.

74. Dinar, S. (2002). "Water, Security, Conflict and Cooperation." *SAIS Review: A Journal of International Affairs* 22 (2): 229–254 (Summer-Fall).

75. See Schweizer, P. (1990). "The Spigot Strategy." *New York Times.* op-ed. November 11.

BIBLIOGRAPHY

Alger, C. F. (ed.) (1998). *The Future of the United Nations System: Potential for the Twenty-first Century.* Tokyo: United Nations University Press.

Barnet, R. & Cavanagh, J. (1994). *Global Dreams. Imperial Corporations and the New World Order.* New York: Simon & Schuster.

Benvenisti, E. (1996). "Collective Action in the Utilization of Shared Freshwater: The Challenges of International Water Resources Law." *American Journal of International Law* 90: 384–415.

Brown Weiss, E. (ed.) (1992). *Environmental Change and International Law: New Challenges and Dimensions.* Oxford: United Nations University Press.

Cavanagh, J. et al. (eds.) (1994). *Beyond Bretton Woods. Alternatives to Global Economic Order.* London: Zed Books.

Chatterjee, P. & Finger, M. (1994). *The Earth Brokers. Power, Politics and World Development.* London: Routledge.

Chossoudovsky, M. (1997). *The Globalisation of Poverty. Impacts of IMF and World Bank Reforms.* London: Zed Books.

Clapp, J. (1998). "The Privatization of Global Environmental Governance: ISO 14'000 and the Developing World." *Global Governance* 4: 295–316.

Commission on Global Governance (1995). *Our Global Neighbourhood.* Oxford: Oxford University Press.

Czempiel, E. O. (1992). "Governance and Democratization." In Rosenau, J. N., & Czempiel, E.-O. (eds.). *Governance without Government: Order and Change in World Politics*. Cambridge: Cambridge University Press: 250–271.

Dominicé, C. & Varfis C. (1996). "La mise en œuvre du droit international de l'environnement." In Rens, I. (ed.). *Le droit international face à l'ethique et à la politique de l'environnement*. Chêne-Bourg: Georg Editeur-collection SEBES: 151–172.

Elliott, L. (1998). *The Global Politics of the Environment*. London: Macmillan.

Finger, M. (1991). "The Military, the Nation-State and the Environment." *Ecologist* 21 (5): 220–225.

Finger, M. & Kilcoyne, J. (1997). "Why Transnational Corporations are Organizing to 'Save the Global Environment'." *Ecologist* 27 (4): 138–142.

Finger, M. & Asún, J. (2001). *Learning Our Way Out. Adult Education at a Crossroads*. London: Zed Books.

Finger, M. & Allouche, J. (2001). *Water Privatization: Transnational Corporations and the Re-regulation of the Water Industry*. London & New York: E & FN Spon.

Ginther, K. (1995). "Sustainable Development and Good Fovernance: Development and Evolution of Constitutional Orders." In Ginther, K., Denters, E. and de Waart, P. (eds.). *Sustainable Development and Good Governance*. Dordrecht, Netherlands: Martinus Nijhoff: 150–164.

Gleick, P. H. (1998). *The World's Water: The Biennial Report on Freshwater Resources, 1998–1999*. Washington DC & Covelo, CA: Island Press.

Habermas, J. (1973). *Legitimationsprobleme im Spätkapitalismus*. Frankfurt, Germany: Suhrkamp.

Hardin, G. J. (1968). "The Tragedy of the Commons." *Science* 162: 1243.

Homer-Dixon, T. (2000). *The Ingenuity Gap. How Can We Solve the Problems of the Future?* New York: Knopf.

Hossain, K. (1995). "Evolving Principles of Sustainable Development and Good Governance." In Ginther, K., Denters, E., de Waart, P. (eds.). *Sustainable Development and Good Governance*. Dordrecht, Netherlands, Martinus Nijhoff: 15–22.

Hunter, D., Salzman, J., & Zaelke, D. (1998). *International Environmental Law and Policy*. New York: Foundation Press.

Jessop, B. (1998). "The Rise of Governance and the Risks of Failure: the Case of Economic Development." *International Social Science Journal* 155: 29–45 (March).

Kooiman, J. (ed.) (1993). *Modern Governance: New Government Society Interactions*. London: Sage.

Korten, D. (1995). *When Corporations Rule the World*. West Hartford CT: Kumarian Press.

Krasner, S. (ed.) (1983). *International Regimes*. London: Cornell University Press.

Kreps, D. M. et al. (1982). "Rational Cooperation in the Finitely Repeated Prisoners' Dilemma," *Journal of Economic Theory* 27: 245–252.

Krishna, R. (1998). "The Evolution and Context of the Bank Policy for Projects on International Waterways." In Salman, M. A. & Chazournes, L. B. (eds.). *International Watercourses. Enhancing Cooperation and Managing Conflict*. Proceedings of a World Bank Seminar, World Bank Technical Paper 414, Washington DC.

Lipschutz, R. (1996). *Global Civil Society and Global Environmental Governance*. Albany: State University of New York Press.

Lipschutz, R. (1997). "From Place to Planet: Local Knowledge and Global Environmental Governance." *Global Governance* 3: 83–102.

March, J. & Olsen, J. (1989). *Rediscovering Institutions. The Organizational Basis of Politics*. New York: Free Press.

McCaffrey, S. C. (1998). "The UN Convention on the Law of the Non-Navigational Uses of International Watercourses: Prospects and Pitfalls." In Salman, M. A. & Chazournes, L. B. (eds.). *International Watercourses. Enhancing Cooperation and Managing Conflict*. Proceedings of a World Bank Seminar. World Bank Technical Paper 414, Washington DC.

Mische, P. (1992). "National Sovereignty and Environmental Law." In Bilderbeek, S. (ed.). *Biodiversity and International Law. The Effectiveness of International Environmental Law*. Amsterdam: IOS Press: 105–114.

Ohmae, K. (1995). *The End of the Nation-State*. New York: Free Press.

Ostrom, E. (1990). *Governing the Commons. The Evolution of Institutions for Collective Action*. Cambridge: Cambridge University Press.

Paolini, A.J., Jarvis, A., & Reus-Smith, C. (eds.) (1998). *Between Sovereignty and Global Governance. The United Nations, the State and Civil Society*. New York: St. Martin's Press.

Peters, G. & Savoie, D. J. (eds.) (1993). *Governance in a Changing Environment*. Toronto: Canadian Center for Management Development.

Postel, S. L., Daily, G. C. & Ehrlich, P. R. (1996). "Human Appropriation of Renewable Fresh Water" *Science* 271, February 9.

Powell, W. & DiMaggio, P. (eds.) (1991). *The New Institutionalism in Organizational Analysis*. Chicago: University of Chicago Press.

Prakash, A. & Hart, J. (eds.) (1999). *Globalization and Governance*. London: Routledge.

Princen, T. & Finger, M. (eds.) (1994). *Environmental NGOs in World Politics: Linking the Local to the Global*. London: Routledge.

Putman, R. D. (1998). "Diplomacy and Domestic Politics: The Logic of Two-Level Games." *International Organization* 42: 427–460.

Rosegrant, M. W. (1995). "Dealing with Water Scarcity in the Next Century." *2020 Vision Brief 21*. http://www.cgiar.org/ifpri/2020/briefs/number21.htm.

Rosenau, J. (1990). *Turbulence in World Politics. A Theory of Change and Continuity*. Princeton: Princeton University Press.

Rosenau, J. (1995). "Governance in the Twenty-First Century." *Global Governance* 1: 13–43.

Rosenau, J. (1997). *Along the Domestic-Foreign Frontier. Exploring Governance in a Turbulent World*. Cambridge: Cambridge University Press.

Sands, P. (1994). "International Law in the Field of Sustainable Development." *British Yearbook of International Law:* 303–381.

Sands, P. (1996). "Compliance with International Environmental Obligations: Existing International Legal Arrangements." In Cameron, J., Werksman, J., & Roderick, P. (eds.). *Improving Compliance with International Environmental Law*. London: Earthscan.

Senarclens, P. (1998). "Governance and the Crisis in the International Mechanisms of Regulation." *International Social Science Journal* 155.

Shihata, I. (1994). *The World Bank Inspection Panel*. Published for the World Bank. Oxford: Oxford University Press.

Stoker, G. (1998). "Governance as Theory: Five Propositions." *International Social Science Journal* 155: 17–28 (March).

Tamiotti, L. & Finger, M. (2001). "Environmental Organizations: Changing Roles and Functions in Global Politics." *Global Environmental Politics* 1 (1): 56–76 .

Thurow, L. (1996). *The Future of Capitalism. How Today's Economic Forces Shape Tomorrow's World*. New York: Morrow.

United Nations, Commission on Sustainable Development (1997). "Comprehensive Assessment of the Freshwater Resources of the World." Report of the Secretary-General. U.N. Doc. E/CN.17/1997/9 (February 4).

Väyrynen, R. (ed.) (1999). *Globalization and Global Governance*. New York: Rowman and Littlefield.

Victor, D., Raustiala, K., & Skolnikoff, E. (eds.) (1998). *The Implementation and Effectiveness of International Environmental Commitments. Theory and Practice*. Cambridge (Massachusetts) and London (England): IIASA/MIT Press.

World Bank (1992). *Governance and Development*. Washington DC: World Bank.

World Bank (1994). *Projects on International Waterways*. World Bank Operational Manual. Operational Policies, OP 7.50.

World Bank (1997). *World Development Report 1997: The State in a Changing World*. Oxford NY: Oxford University Press.

World Water Council (1999). *World Water Vision: Report*. World Water Commission's Staff Report. Cosgrove, W. J., & Rijsberman, F. R. (eds.). Version of December 17.

Young, O. (1989). *International Cooperation: Building Regimes for Natural Resources and the Environment*. Ithaca NY: Cornell University Press.

Young, O. (1994). *International Governance. Protecting the Environment in a Stateless Society*. Ithaca NY: Cornell University Press.

Young, O. (1995). *Global Governance. Drawing Insights from the Environmental Experience*. Hanover NH: Dartmouth College.

Chapter 2

The Mekong River Basin: Comprehensive Water Governance

Nantana Gajaseni, Oliver William Heal,
and Gareth Edwards-Jones

The Mekong River Basin is an area of natural wealth and diversity nourished by the Mekong River and by its tributaries. As the river flows through the Indo-China region through six countries, it carries with it a rich silt-load that is deposited on fertile flood plains. The basin is home for some sixty-six million people and to a massive array of terrestrial and aquatic life. Farming and fishing are the ordinary local livelihood activities, which are closely linked to the forest and river ecosystems and sustain daily life. With the indigenous knowledge and the relationships to the environment, the people know how to deal with, adapt to, and survive in the natural condition. Comprehensive water governance[1] in this basin is very important for global, regional, and local communities. It requires transparency of water resource management policy, effective public resource management, and a stable and regulatory environment for sustainable society and the ecosystem. Good comprehensive governance can improve the water resource management based on the concept of a holistic approach for conservation and sustainable resource use and for equitable sharing of the benefits at all levels. To achieve this approach, it is essential to provide all stakeholders with balanced, objective information on the consequences of different options for water management.

I. BACKGROUND

1.1. The Characteristics of the Mekong River Basin

1.1.1. Physical Characteristics

The Mekong River Basin, the twenty-first largest basin in the world, is only one of the major international rivers in the Indo-China region, which flows through six countries; China, Myanmar, Lao PDR, Thailand, Cambodia, and Vietnam. The river flows for approximately 4,800 km, the twelfth rank in the world, from the origin on the northeast rim of the Tibetan plateau to the South China Sea through the Mekong Delta of Vietnam (Figure 3). Division of the Mekong River Basin by countries creates artificial divides. The natural division of the basin is both by subwatershed and by a biogeographic zone, which is divided into 'Upper Mekong Basin' and 'Lower Mekong Basin.' The Upper Mekong Basin is located in Yunnan, China, and in the east of Myanmar. The Lower Mekong Basin is located in Lao PDR, Thailand, Cambodia and Vietnam.

The hydrologic regime of the river is primarily dependent on the monsoon climatic conditions, with mean annual precipitation in the Lower Mekong Basin of 1,672 mm.[2] Water flow and flooding in the rainy season produces a total runoff of 475,000,000 cu m annually.[3] The high-water period is from August to October and the low-water period is from February to April. The flood season in the mainstream and tributaries, from June/July to November/December accounts for 85–90 percent of the total annual water volume. With the high total volume of runoff sustaining this basin, it is necessary to seek adequate governance for water resource management among the riparian countries.

Owing to the different landforms of the catchment, the Mekong River Basin is divided into six sensitive geographic areas (Table 3), with each landform influenced by different annual rainfall. The surface water runoff from the catchment and through the river and tributaries, regularly supplies nutrients during the flood period to the lowland and to the delta.

TABLE 3. The characteristics of the main landform of the Mekong River Basin

Landforms	Area covered	Rainfall (mm/year)
1. Lanchang River Basin	Yunnan, China	600–2,700
2. Northern Highlands	East Myanmar, North Thailand, North Lao PDR	2,000–2,800
3. Korat Plateau	Northeast Thailand, Southwest Lao PDR	1,000–1,600

(continued)

TABLE 3. The characteristics of the main landform
of the Mekong River Basin *(continued)*

Landforms	Area covered	Rainfall (mm/year)
4. Eastern Highlands	Southeast Lao PDR, Southwest Vietnam	2,000–3,000
5. Lowlands	South Lao PDR, North Cambodia, South Vietnam	1,100–2,400
6. Southern Uplands	Southwest Cambodia	Up to 4,000

MRC, 1996.

The physical characteristics of the basin provide potential as a major hydropower energy source, with development built on the mainstream to improve the economies of the riparian countries. Energy demand in the Mekong River Basin, especially in Thailand and Vietnam, is increasing at a very high rate. The aggregated growth projected in the six riparian countries in the ten-year period 1993–2003 is more than 12,000 MW generating capacity and 120,000 GWh/y energy generation.[4]

River transport is also important in various parts of the basin between Yunnan to the South China Sea. Navigation is efficient in some, for instance from Yunnan to Lao PDR, with domestic transport in Cambodia and the delta and in the tributaries from the South China Sea to Phnom Penh.[5] Uninterrupted navigation from Yunnan to the delta is not feasible due to physical conditions such as rapids, shoals, and sharp bends, which restructure communication among the riparian countries. Improved navigation use is an attractive option for development projects.

1.1.2. Biological Characteristics

Due to the overlap of distribution of fauna in the Himalayas[6] and Sunda[7] zoogeographic zones, all of the countries in the Mekong River Basin show high genetic, species, and ecosystem diversity. The fauna from the southern oriental region in Sunda zone is distributed northward to the Southeast Asia continent, while the fauna from Himalayas zone grades southward to the Indo-China subregion. Moreover, species of both zones cohabit in the Mekong River Basin, and areas with particularly rich biodiversity are located along the national borders because these areas are inaccessible and remote from human settlement and transportation links. Biodiversity in the Mekong River Basin, includes various types of terrestrial, wetland and aquatic ecosystems. The list of fauna and flora recorded in the Mekong River Basin indicates that the basin is very rich in biodiversity at the 'species level' (Table 4). It is known that a certain number of species have been exterminated, and a

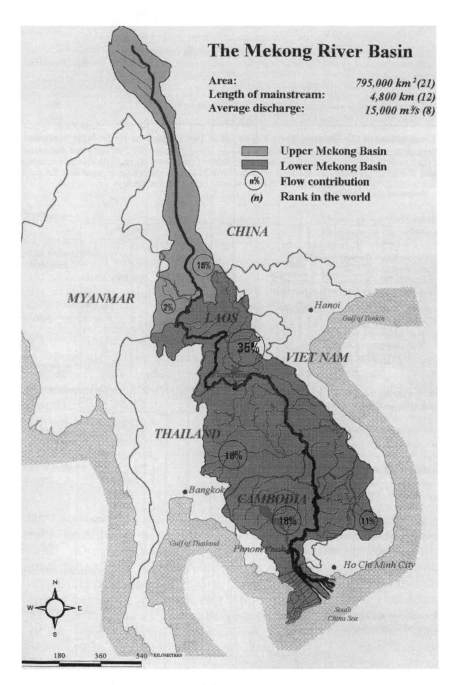

FIGURE 3. General characteristics of the Mekong River Basin

number of species have probably not been discovered yet. The areas with the richest biodiversity and hot spots as the habitats of rare species are the border triangle of Cambodia, Lao PDR, and Vietnam, the border of Cambodia and Thailand, and the quadrangle of Yunnan, Myanmar, Lao PDR, and Thailand.[8]

TABLE 4. The list of fauna and flora in the Mekong River Basin. The apparent lower diversity in Lao PDR and Cambodia probably reflects a lower intensity of research.

Feature	Yunan	Myanmar	Lao PDR	Thailand	Cambodia	Vietnam
Mammals (Total)	255	300	157	282	117	275
Mammals (Endemic)	N/A	6	1	8	1	5
Birds (Total)	766	1,000	609	930	545	744
Birds (Endemic)	N/A	3	3	2	0	4
Freshwater fishes	N/A	N/A	N/A	650	850+	N/A
Amphibians (Total)	N/A	75	37	107	28	80
Amphibians (Endemic)	N/A	N/A	N/A	13	N/A	N/A
Reptiles (Total)	N/A	360	66	298	82	180
Reptiles (Endemic)	N/A	N/A	N/A	31	N/A	N/A
Swallowtail butterflies	N/A	68	39	56	22	37
Insects	N/A	N/A	N/A	N/A	N/A	6,000
Vascular plants (Total)	18,000	7,000	8,290	15,000	7,570	12,000
Vascular plants (Endemic)	N/A	1,071	1,457	2,742	1,175	4,800
Ferns	N/A	N/A	N/A	600	N/A	800
Fungi	N/A	N/A	N/A	3,000	N/A	600
Total of biodiversity	19,021	9,877	10,655	23,665	10,389	25,516

Note: N/A is not available.
MRC, 1996

In the Mekong River Basin, there are three major types of ecosystems with different characteristics—terrestrial, wetland, and aquatic ecosystems—which can be described as follows:

Terrestrial ecosystems in the Mekong River Basin are dominated by tropical forests (evergreen, deciduous, and dipterocarp forests) which represent the productivity of the watershed catchment areas. The decline of forest cover with increasing rate of deforestation is indicated by the forestry data for the 1990–1995 period in Table 5.

TABLE 5. Forestry statistical data in the riparian countries
in the Mekong River Basin

Description	Myanmar	Lao PDR	Thailand	Cambodia	Vietnam
Total land area (1000ha)	67,658	23,680	51,312	18,104	33,169
Forest area 1990 (1000ha)	29,088	13,177	13,277	10,649	9,793
Forest area 1995 (1000ha)	27,151	12,435	11,630	9,830	9,117
% forest 1995	40.1	52.5	22.7	54.3	27.5
Forest change 1990-95 (1000ha)	–1,937	–742	–1,647	–819	–676
Annual change rate 1990-95 (%)	–1.4	–1.2	–2.6	–1.6	–1.4

FAO, 1999 and WRI, 1999.

Wetland ecosystems in the Mekong River Basin are intimately linked to ecological balance and economic well-being. They are the major habitats for fish, which is the major protein-source for the basin's inhabitants. The wetland ecosystems are also significant as wildlife sanctuaries for birds, amphibia, reptiles, etc. The population dynamics of the aquatic plants and animals of wetland ecosystems (including shrimp/ prawn, crayfish, and fish) revolve around the seasonal ebb and flow of the monsoon rains and floodwaters. For instance, mangrove forest and *Melaleuca* forest[9] are wetland ecosystems in the Mekong Delta in Vietnam that have high fishery productivity. Because of defoliation during the war; deforestation due to logging; firewood collection; and increase in rice and shrimp farming, the wetland forest has been reduced from the original 500,000 ha to 233,200 ha in 1988.[10]

Aquatic ecosystems in the Mekong River Basin represent significant biological value in terms of species composition and diversity. The Mekong River Commission (MRC) reports that there are twelve hundred different fish species in this basin.[11] The Mekong River Basin produces as much as 1 million tons of fish annually.[12] For instance, two groups of freshwater fish are distinguished based on their spawning behavior and environmental tolerance, White fish (*Cyprinidae* and *Schilbeidae*) and Black fish (*Clariidae, Siluridae,* and *Ophiocephalidae*). Moreover, in Mekong River, there is the Giant catfish (*Pangasinodon gigas*), which is the world's largest freshwater fish and commonly weighs 150–200 kg.[13] There is also the largest freshwater lake in Southeast Asia called 'Tonle Sap' or 'Great Lake' in Cambodia, which is a productive fishing ground. Eight hundred fifty fish species have been identified from this lake and from the river mainstream.[14]

In addition to its high economic benefits, the basin is of considerable value in terms of conservation. Effective management is necessary in order to sustain high productivity for the local and regional people and also for global conservation of biodiversity.

The Historical Background of the Mekong River Basin

Prior to 1975, the United States influenced this region during the Vietnam War. The U.S. government withdrew military support from Vietnam in 1973; the communist regime took control of the country afterward. After 1975, the communist victories in Cambodia, Lao PDR, and Vietnam signaled a fundamental geopolitical shift; a 'Cold War division,' rather than integration within the region ensued due to a tense policial environment. The relationships between Thailand and Lao PDR were tense throughout the 1980s; the bloody conflict in early 1988 was over border demarcation affecting three villages on the border of Uttaradit and Sayabouri provinces. During these periods of war, the ecological, social, and economic situation in the basin was devastated.

Until 1991, the Paris peace agreement signaled an end to full-scale conflict in Cambodia, and the reduction in tension and revived interests in developing the basin's resources. Five of these countries have been through some form of socialist or communist administration, the exception being Thailand, and three (China, Lao PDR, and Vietnam) remain nominally socialist states in the process of economic reform.[15] Although Thailand has been a democracy for a long time, the system is still fragile, and there was little development of the Mekong River's resources until 1995, when the Agreement on Cooperation for the Sustainable Development of the Mekong River Basin was signed in Chiangrai, Thailand. This opened the way for resumption of cooperation between four riparian countries with strong recommendations for China and Myanmar to join the Mekong River Commission. Apparently, the future of the environmental, social, and economic situations in the Mekong River Basin is optimistic regarding sustainable development.

1.3 The Stakes of the Mekong River Basin

1.3.1. Geopolitical Situation

The Mekong Basin is divided along national lines into six riparian countries. The cumulative effects of geopolitical tensions due to wars and communist regimes in the basin have limited resource development. As a result, the Mekong River is one of the least modified of major world rivers in terms of impoundments and diversion, particularly in the case of the mainstream and tributaries. Tensions have emerged on the issue of water sharing in relation to countries' rights of unilateral action or alternatively rights of veto, which are essential issues in regional cooperation for water resource development and management.

For instance, Yunnan Province is located in the Upper Mekong Basin, where river and catchment changes have significant implications for downstream resource developments. There are few indications of China's perspective on downstream impacts of developments in its territory, and for negative impacts upstream of the sequence of dam constructions in Yunnan. This makes minimization of potential downstream impacts nonnegotiable and does not facilitate a regional cooperative approach to basin management. The downstream riparian countries are left in a powerless position. Even the Mekong River Commission cannot enforce or negotiate with China regarding water resource development and management. As a result there are conflicts on equal water sharing and water benefit among the riparian countries.

Due to the characteristics of the basin as a rich cultural heritage and diversity, for instance, hundreds of different ethnic groups—with their own language, traditions, and beliefs—live in the lowland, upland, and highland communities throughout the region. The Mekong River is a prosperous river, which contributes its natural resources to not only humanity but also to other terrestrial and aquatic life in this region. Within a total land area of 795,000 sq km in the Mekong River Basin, many farmers and fishermen depend on the natural resource base to sustain their livelihood.[16] Various ethnic groups reside in mountainous and undeveloped areas:

- Jarai and Ede (central highlands and Vietnam) practice sustainable harvesting of different fish species based on their particular spawning cycle.
- Karen (northern Thailand) regulate the use of timber and biological resources in the village vicinity.
- Hani (Yunan) have sustainable food production through swidden cultivation and home gardens.
- Lao Loum (lowland), Lao Theung (upland), and Lao Suung (highland) are the majority ethnic groups in Lao PDR, which basically use cultivation practices.

Most ethnic groups have no conflicts in terms of natural resource use. However, the ethnic minority groups Wa and Shan in Shan State, Myanmar, have conflicts with the government in relation to drug production and trafficking from the Golden Triangle between Myanmar, Lao PDR, and Thailand.

1.3.2. Environmental Situation

The environmental situation can be characterized in terms of quality. The soil quality and productivity in the basin are affected by natural processes (e.g., wind, leaching of salts and acids, and saltwater intrusion) and by human impacts (e.g., intensive farming, and deforestation). The erosion rates are high in the rainy season (e.g., in the central highlands the rates reach 150–170 ton/ha/year on slopes of 20–22°C) and low in the dry season. Erosion is assessed as moderate severity in the region.[17]

The Mekong's water quality is strongly related to the flow rate and to the local physiography. Natural processes of leaching of salts, acids, and other organic matter and nutrients from soils, saltwater intrusion, drainage runoff, evapotranspiration, suspended solid deposition, and the weathering of bedrock minerals determine the rates of dilution and components of the water. Water quality throughout the Mekong River Basin is generally good, but some localized areas have moderate to severe levels of eutrophication, nutrient loss, organic pollution, salinization, acidification, toxic metals, and microbial pollution.[18]

The Mekong's total suspended solids concentration is about 300 mg/l, compared to the Ganges River in India at 1,130 mg/l.[19] However, the Mekong River has high phytoplankton productivity, averaging 116 $\pm$ 42 mgC/m^2/h (1.05–2.12 gC/m^2/d), which supports important food chains and that provides the high nutrient supply to the wetland ecosystem in the Mekong Delta in Vietnam.[20]

Human impact on the environment is mainly through hydropower dam development. For example, the first three dams in Yunnan have resulted in decreasing water flow in the rainy season and increasing water flow in the dry season. The negative impact is the reduction of inflow to Tonle Sap (Cambodia) during the rainy season, which potentially reduces fish production. These hydrological and ecological changes continue to be a main issue for the four riparian countries of the Mekong River Commission.[21]

Another artificial environmental problem, the intensification of agriculture in the Mekong Delta, is reliant on increased use of agrochemicals such as carbamate and organophosphate pesticides—with shorter residence times in the environment—and organochlorine pesticides—with long residence times in the environment. In a 1996 report, the Mekong River Commission reported that the organic pollution is severe to moderate in the development area and the organochloride pesticides contamination shows moderate severity in the region.[22]

A. Grainger indicated that the causes of tropical deforestation are related more to land use problems than to forest management.[23] The causes in the Mekong Basin are shifting cultivation, excessive or inefficient commercial logging, land encroachment for human settlements, farming and infrastructure development, and heavy fuel wood use. These activities are complicated by uncertain land ownership.[24] The habitat loss—terrestrial, wetland, and aquatic—caused by the rapid population growth and increased resource consumption has caused loss of wildlife in particular, endangered species such as Douc monkey (*Pygathrix nemaeus*), Black gibbon (*Hylobates concolor*), Asian elephant (*Elephas maximas*), Less adjutant stork *(Leptoptilos javanicus)*, and Milky stork *(Mycteria inerea)*.[25] The *Melaleuca* forest in the wetland ecosystem was devastated during the Vietnam War by the chemical defoliants napalm and bombing combining with the pressure of drainage for rice cultivation. This forest originally covered half of the 1.8 million ha of acid-sulphate soils but the Vietnam War reduced the forest area to 174,000 ha by 1972.[26] The loss of *Melaleuca* forest resulted in economic and ecological changes,[27] with increased natural water quality problems, particularly salt and acid-soil leaching and saltwater intrusion.

In relation to global climate changes, the high rate of tropical deforestation in the basin of 1.2–2.6 percent annually[28] contributes to increasing carbon dioxide emission to the atmosphere from this basin. Thus forests are changing their roles from carbon sink to become carbon source. If the forests are protected, they are capable of absorbing atmospheric carbon dioxide and generating a carbon sink in biomass at 100 tonneC/ha in the primary dry dipterocarp forest. The economic value of the existing primary forest in the Lower Mekong River Basin is estimated to be in the range of $1,100–20,500 ha based on carbon sequestration in tree biomass.[29] This carbon sequestration evaluation can be considered another economic benefit from the Mekong River Basin at the ecosystem diversity level and is of global relevance. Protection of the forest will also affect water quality, with potential reduction of soil erosion and reduction in chemical pollution.

1.3.3. Socioeconomic Characteristics of the Basin

Due to the historical differences and divisions within the basin, the main economic situation is notable in the different levels of income between the wealthiest and poorest countries (Table 6). Social indicators such as the percentage of population below the national poverty line, the life expectancy at birth, and the percentage of illiteracy, conclude that the people of most of the riparian countries are poor, with the exception of Thailand. The population in the riparian countries in 2000 and 2025 based on projections from 1990 base data indicate a population increase.[30] Concerning the economic situation of the whole basin, the Gross National Product (GNP) per capita in 1998 was highest in Thailand at $2,200 and lowest in Cambodia at $280. These riparian countries have a mostly poor economy when compared to the GNP per capita in East Asia and Pacific countries ($990) and in the low-income countries ($520).[31] The people in each riparian country depend on natural resources and the greatest percentage of water use is in the agricultural sector. The industrial and domestic sectors are simply related to the proportion of water supply per capita from Mekong River as a renewable freshwater resource.

TABLE 6. Socioeconomic and natural resource indicators of the riparian countries, parts of which form the Mekong Basin

Description	Myanmar	Lao PDR	Thailand	Cambodia	Vietnam
Population 1998 (million)	44.4	5.0	61.1	10.7	76.7
Population 2000* (million)	45.6	5.4	61.4	11.2	79.8
Population 2025* (million)	58.1	9.7	72.7	16.5	108.0
Average annual growth rate (1992–98, %)	1.2	2.6	1.1	2.6	1.7
Poverty (1992–98, % below poverty line)	N/A	46	13	36	37
Life expectancy at birth (1992–98, years)	60	53	69	54	68
Illiteracy (%of population age 15+, 1992–98)	16	N/A	5	N/A	17
GDP 1998 (US$ billions)	N/A	1.6	111.3	3.0	25.9
GNP per capita 1998 (US$)	N/A	330	2,200	280	330
Renewable freshwater 2000*(1000m³/cap)	19.3	35.0	3.4	10.8	4.6
Water use %* (domestic/industry/agriculture)	7/3/90	8/10/82	5/4/91	5/1/94	4/10/86

World Bank, 1999 and Dore, 2001.

Based on the distinguishing characteristics of the Mekong River, there are attempts to utilize the hydroenergy and to improve navigation for the development in the Indo-China region. China places an emphasis on navigation to interact with downstream nations and the MRC has carried out this strategy for developing upper Mekong navigation between Lancang Jiang and the Mekong River in order to boost the economy.[32] The cost-effective link between China and three adjacent countries (Myanmar, Lao PDR, and Thailand) has led to greatly increased traffic on the Mekong River, and other improvements. It is necessary to restructure by blasting rapids and shoals with dynamite, to increase the year-round capacity of the route to accommodate 100–300-ton vessels.[33] This kind of shipping development may change ecological characteristics and impact biodiversity. However, the potential of hydropower dam development in the Mekong River Basin is likely to be of greater importance; the existing and the proposed dams are identified in Table 7.

TABLE 7. Proposed and existing hydropower dam developments in the Mekong River Basin

Hydropower dam	Country	Capacity (MW)	Source
Upper Mekong Basin			
Manwan Dam*	China	1,500	Chapman and Daming, 1996
Dachaoshan Dam	China	1,350	Chapman and Daming, 1996
Jinghong Dam	China	1,500	Chapman and Daming, 1996
Xiaowan Dam	China	4,200	Chapman and Daming, 1996
Nuozhadu Dam	China	5,000	Chapman and Daming, 1996
Mengsong Dam	China	600	Chapman and Daming, 1996
Gonguoqiao Dam	China	750	Chapman and Daming, 1996
Lower Mekong Basin			
Nam Ngum Dam	Lao PDR	150	Hirsch and Cheong, 1996
Nam Theun-Hinboun Dam	Lao PDR	210	IRN, 1999
Nam Luek Dam	Lao PDR	60	IRN, 1999
Pak Beng Dam*	Lao PDR	1,410	MRC, 1994
Luang Prabang Dam*	Lao PDR	2,560	MRC, 1994
Sayaburi Dam*	Lao PDR	1,260	MRC, 1994
Pak Lay Dam*	Lao PDR	1,760	MRC, 1994
Pa Mong Dam*	Thailand & Lao PDR	2,440	MRC, 1994

(continued)

TABLE 7. Proposed and existing hydropower dam developments
in the Mekong River Basin *(continued)*

Hydropower dam	Country	Capacity (MW)	Source
Ban Koum Dam˚	Thailand & Lao PDR	2,700	MRC, 1994
Pak Mun Dam	Thailand	136	IRN, 1999
Don Sahong Dam˚	Lao PDR & Cambodia	240	MRC, 1994
Stung Treng Dam˚	Cambodia	1,190	MRC, 1994
Sambor Dam˚	Cambodia	4,110	MRC, 1994
Yali Falls Dam	Vietnam	700	Hirsch and Cheong, 1996

* Mainstream dams   bf = Proposed dam developments

The 1994 Run-of-River Hydropower Plan on the mainstream of the Lower Mekong River was premised by the need to select sites to minimize socioeconomic and environmental impact.[34] The total energy-generating capacity of the nine proposed dams is about 13,350 MW.[35] Electricity from the mainstream hydropower dams will provide low-cost energy to improve the energy supply situation in the region. Hydropower energy is usually considered cheap compared to energy from other sources (oil and coal) and also have implications for global climate change.

The nine proposed hydropower dams on the mainstream of the Lower Mekong River would displace an estimated 61,200 people and increase land pressures in resettlement areas.[36] Agriculture would also be affected if the dams reduce or eliminate the nutrient-rich silts deposited by floodwaters, and the remaining floodplain soils would be threatened by salinization if reservoirs caused underground salt deposits to dissolve and leach to the soil surface. Therefore in the near future, the local and regional natural resources are likely to be further degraded if the developments in the Mekong River Basin are implemented without a holistic assessment and comprehensive governance of resource management.

Meanwhile, the international finance institutions search for profitable resource development opportunities and riparian countries are under extraordinary pressure for rapid economic development with high natural resource utilization. This has major implications for resource management and countries need to reform governance in order to ensure effective long-term natural resource use to improve their economies.

2. ANALYSIS OF RELEVANT INSTITUTIONS AND ACTORS

2.1. *Identification of the Relevant Institutions and Actors*

2.1.1. Global Level

The multi-lateral or international institutions involve the resource management in this basin, which include the following:

- The UN with major environmental responsibilities including[37]
  - Food and Agricultural Organization (sustainable agricultural and forestry)
  - International Fund for Agricultural Development (sustainable agriculture)
  - United Nations Center on Transnational Corporations (sustainable development)
  - United Nations Commission for Sustainable Development (implementation of Agenda 21)
  - United Nations Development Programme (sustainable development)
  - United Nations Education, Scientific, and Cultural Organisation (environmental monitoring-Man and the Biosphere Programme)
  - United Nations Environment Programme (environmental policy coordination)
  - World Food Programme (sustainable agriculture)
- World Bank
- Asian Development Bank (ADB)
- Global Environmental Facility (GBF)
- Japan Bank for International Co-operation (JBIC)

For instance,

- UNDP has supported the Mekong River Commission to focus on capacity building with expected outcome as "a draft Programme Support Document (PSD)" for a three-year programme of assistance. This programme will ensure that the implementation of the 1995 Mekong Agreement on Sustainable Development of the Mekong River Basin is activated and worked out properly.[38]
- The World Bank has approved financial funds from the Global Environment Facility (GEF) to the Mekong River Commission of $11 million since February 2000. This grant will support the MRC and the member states to ensure that development of the water resources is carried out in a sustainable manner that preserves the environment. Implementation of the Mekong Agreement requires strong political commitment from all member states and the participation and support of stakeholders in the basin and external parties.

These multi-lateral institutions play very important roles representing the specific interests and activities of the global society, with differing perspectives, goals, and purposes. The diverse interests of these bodies, range through political, social, economic, and environmental aspects in relation to specific project issues and programmes. The national and regional governments and organizations therefore have real problems of integrating the differing views.

2.2.2. Regional Level

The Mekong River is an international river flowing through six countries. Four riparian countries (Lao PDR, Thailand, Cambodia, and Vietnam) established an international organization, the Mekong Committee in 1957, to address the comprehensive development of water and related resources in the Lower Mekong Basin. Funding came from the UN's regional organization, the Economic Commission for Asia and the Far East (ECAFE), a precursor to the Economic and Social Commission for Asia and Pacific (ESCAP). The Mekong Committee was seen as a catalyst for development of the Mekong Basin's resources to increase the level of income for the riparian countries.[39]

The Interim Mekong Committee was formed in 1978 after Cambodia dropped out in 1975. The purpose of this committee is to coordinate the work of the riparian countries in order to maximize the social and economic benefits by following a course of sustainable development of the Mekong's water resources.

The four lower riparian countries established the Mekong River Commission in April 1995 with the signing of the Agreement on Cooperation for the Sustainable Development of the Mekong River Basin. This institution lies in part at the regional level, seen most clearly in the precursors, but also at the national level as agency roles have changed and developed. Its role is very important in policy planning for sustainable development, utilization, management, and conservation of water and related resources of the Mekong River Basin. The MRC depends on the resources from member countries, international donor communities, and cooperating agencies.

The MRC has three permanent bodies, namely (1) the council (at the minister/cabinet levels); (2) the joint committee (at the department head level); and (3) the secretariat (the technical and administrative arm of the commission). The mandate of the MRC is

- To cooperate and promote in a constructive and mutually beneficial manner in the sustainable development, utilization, conservation, and management of the Mekong River Basin.
- To protect the environment, natural resources, aquatic life and conditions, and ecological balance of the Mekong River Basin from

pollution or from other harmful effects resulting from any development plans and uses of water and related resources in the Basin.[40]
- To define as the 'general planning tool and process that the Joint
  Committee would use as a blueprint to identify, categorise and
  prioritise the projects and programmes to seek assistance for and
  to implement the plan at the basin level.'

According to Yasunobu Matoba, Chief executive officer, The
Mekong River Commission's Secretariat, "In developing and using water
resources, priority has to be given to the satisfaction of basic needs and
the safeguarding of ecosystems."

Recently, the national policies of riparian countries have been guided
and focused on sustainable development of the Mekong River Basin
since the MRC Agreement of 1995. The agreement has forty-two articles that include key articles relate to requirements on all members to
notify and consult concerning projects on the tributaries and mainstream of the Mekong River. It aims to develop rules for water utilization and interbasin diversions, i.e., water sharing (Art. 26) and to
formulate a basin development plan, really a planning process (Art. 24).
It also very clearly specifies the functions of each component of its
institutional framework (Arts. 15–33) and the process for dispute resolution (Arts. 34–35).[41] The agreement reflected a positive interest and
opportunities to establish cooperation for a sustainable future in the
basin. Of significance has been the creation of new national environment bodies in the early 1990s. These include

- Ministry of Science, Technology and Environment in Vietnam (pre-
  viously State Committee for Science and Technology)
- Ministry of Environment in Cambodia (new entity)
- Science, Technology and Environment National Organisation in Lao
  PDR (previously Science and Technology National Organisation)
- Ministry of Science, Technology and Environment in Thailand
  (previously National Environment Board)

Other agencies responsible for environment and resource management have gone through significant changes including ministries, de-
partments, and agencies dealing with forestry, fisheries, agriculture,
irrigation, energy, and mineral resources. Various international group-
ings, apart from the MRC, both formal and informal, also lend a regional perspective to the Mekong Basin.

Despite the lack of comprehensive governance among the riparian
countries, hydropower development projects seem to be flourishing as
indicated by the support of international organization (Table 7). Six
proposed hydropower projects in the Upper Basin and nine in the Lower

Basin, reflect the high priority given to water resource utilization in the whole basin. If the water resource management policies or plans considered only economic benefits to the government, they might create an imbalance between the benefits to the environment, biodiversity, social communities, and local economic situations. Therefore, it is essential that various alternatives are explored in order to consider balances between the three dimensions of sustainable development (social, economic, and environmental).

2.1.2. Local Level

People are beginning to assert their right to participate in their own governance and they have become important actors in water resource management. Increasing population and economic growth have placed additional pressures on natural resources and on the environment. Management of both demographic and economic change to safeguard the interests of future generations has become an issue of importance, particularly for people in the Mekong River Basin. Empowerment of people is reflected in the vigor of civil society and in the democratic process, and must be vital to meet the environmental, social, and economic challenges. The vigorous key of civil society reflects a large increase in the capacity and will of people to take control of their own lives and to improve or transform them. Empowerment will not be sustained if people lack a stable income or are poor. The socioeconomic-political situation is weak in most of the riparian countries, especially Myanmar, Lao PDR, and Cambodia. Wider educational facilities, improved opportunities for women, and greater access to information as well as political progress, are needed to improve participation and governance at the local level.

Environmental pressures are linked to poverty. Poor people press on remote areas and forests, overexploiting them to survive and undermining the resource base on which their well-being and survival depend. However, the environmental exploitation has not been caused by the consumption of poor people alone. Natural resources are transferred increasingly from the remote areas to the urban communities rather than used for local consumption. It is important to seek strategies for the comprehensive water governance in this basin at the local level.

Actors at the different levels should include

- States and local government authority at national, provincial, and community levels
- Grassroot as there are about 240 million people in the Mekong Basin

- Active nongovernmental organizations (NGOs) such as voluntary associations, women's groups, religion-based organizations, chambers of commerce, and neighborhood communities
- Media such as mainstream media, independent media, and e-mail networks
- Businesses such as national/local businesses, even transnational corporations (TNCs) and other international businesses, consultants, and private financiers, also the deal arranger and insurers
- Bilaterals and multi-laterals such as the UN, the World Bank, the International Monetary Fund (IMF), ADB, WTO, and JBIC
- Policy research institutes such as the Cambodia Development Resource Institute (CDRI); the Centre for Biodiversity and Indigenous Knowledge (CBIK, Yunnan); the Environment Research Institute (ERI, Laos); Thailand Development Resource Institute (TDRI, Thailand); and the Vietnam Environment and Sustainable Development Centre (VNESDC, Vietnam)
- Universities such as Cantho (Vietnam), Chiang Mai (Thailand), and Sydney (Australia Mekong Resource Centre)
- Research and/or advocacy networks such as the Asia Resource Tenure Network (ARTN), the Development Analysis Network (DAN), the Greater Mekong Sub-region Academic Research Network (GMSARN), the International Centre for Research on Agroforestry (ICRAF), the International Mekong Research Network (IMRN), Oxfam Mekong Initiative (OMI), and the European Association of South-East Asian Studies (EUROSEAS)

The subnational, regional, and international NGOs involved in the resource management of basin include:[42]

- Subnational NGOs:
  Assembly of the Poor (Thailand)
  Northern Development Foundation (Thailand)
  Cambodian NGO Forum (Cambodia)
- Regional NGOs:
  Asean-Pacific Forum for Women Law & Development (APWLD)
  Towards Ecological Recovery and Regional Alliance (TERRA)
  FOCUS on the Global South
- International NGOs:
  Community Aid Abroad (CAA)
  World Vision
  Care International (CARE)
  Mennonite Central Community (MCC)
  Oxfam (United Kingdom, Hong Kong)

Australian Catholic Relief (ACR)
Japan International Volunteer Centre (JVC)
World Conservation Union (IUCN)
Wetland International (WI)

More details of NGOs show the following features:

- CAA's aim is to build a more environmentally sustainable world. The integrated development programs include several activities related to not only educational development and building-community capacity but also to agricultural extension and community fisheries that are conservation based.
- Oxfam (Hong Kong) promotes total human development in sustainable, people-based, gender-fair, and environmentally sound activities by funding and nonfunding support to community-based groups and to local NGOs.
- JVC promotes sustainable development, self-reliance, and the conservation of indigenous culture and natural resources through people-to-people communication and cooperation.
- IUCN's aim is to support the development of adequate capacity within the Mekong Basin for management of natural areas, ecosystems, and species.
- WI's aim is to sustain and restore wetlands, and their resources and biodiversity for future generations through research, information exchange, and conservation activities worldwide.

These international NGOs constantly offer knowledge, skills, enthusiasm, a nonbureaucratic approach, and grassroots perspectives to the local people of riparian countries and also attributes that complement the resources of official agencies. If the local authorities and the NGOs are strong and based on adequate understanding of the goal of their activities, they must retain independence and avoid being portrayed as foreign-influenced. Overall, the local organizations and the NGO sector, together with intergovernmental organizations at the national level, are not yet significantly developed to contribute effectively to comprehensive governance in resource management in the basin. Poverty, poor democratic political systems, and centralized government, obstruct cooperation between local NGOs and local governmental organizations. Furthermore, the absence of established public participation processes in most of the riparian countries reflects the weakness of public influence on the planning, construction, or operation of development projects.[43]

2.2 Relationships and Interconnections between the Different Institutions

The concept of governance can no longer be considered a closed system.[44] The new comprehensive governance is the exercise of political, economic and administrative authority at global, regional, and local levels to sustain resource use.[45] This approach is driven by three sectors: state or government, private sector, and civil society. Government can create a conducive political and legal environment. The private sector can generate economic growth at local and national levels. Civil society facilitates political and social interaction, mobilizing groups to participate in activities related to water utilization in the basin.

The multi-lateral institutions can create closer relationships and interconnections to strengthen the concept of comprehensive water governance within their policies or actions. In addition to the unstability of political systems and low economic growth, the riparian countries of the Mekong River still need assistance in all aspects from these multilateral or international organizations. Assistance is needed to improve and reform the national governance concerning[46]

- the transparency of government accounts
- the effectiveness of public resource management
- the stability and transparency of the economic and regulatory environment for private sector activity

Given commonality of interests and linkages among other multilateral institutions, national governments and the MRC could reform resource management policy and plan toward more comprehensive governance. For instance, IMF could seek to strengthen its collaboration on issues of water resource governance with the riparian countries and, in particular, with the World Bank or ADB.[47] A summary of multi-lateral organization interconnections related to water resource management is given in Table 8.

TABLE 8. Summary of multi-lateral organization interconnections regarding water resource management

Multi-lateral organizations	Interconnections for water resource management
World Bank, UNDP, UNICEF	To improve the efficiency of water usage in water and sanitation programs.
UNEP, UNDP, IFAD, and others	To negotiate responsibilities for implementing development plans arising from the UN Conference on Environment and Development.

(continued)

TABLE 8. Summary of multi-lateral organization interconnections regarding water resource management *(continued)*

Multi-lateral organizations	Interconnections for water resource management
UN organizations	To reform or reorganize development and expedition of new water technology and water extraction and collection projects.
UNEP and others	To be considered for opportunities for internationally tradable rights in countries where demand is different, e.g., industrial vs. agricultural, different level of development, technology, or preferences. It also considers making benefits from Joint Implementation (JI) schemes for CO_2 emission control.
World Bank, UNDP, and UNEP	To establish the 'Global Environmental Facility' for the protection of the global environment in four areas. International waters are one area of protection that will - catalyze scientific and technical analysis, - advance environmental management activities, - ensure the development and management of investment projects, - mobilize private sector resources to support national sustainable development strategies.
World Bank and ADB	To train governments on issues of environmental impact assessment and establish a system of national accounts that includes the economic impacts of the depletion of natural resources.
ADB	To implement policy on governance in 1995 and subsequent diagnostic completed, 1999–2001 for Thailand, Cambodia, Vietnam, and Laos. To support technical assistant (TA) project on specific issues relating to information, monitoring, strategic framework, transborder issues, wetland, etc.

However, it is necessary to activate and strengthen the interconnection of actors not only domestic organizations but also multi-lateral ones in order to create successful governance by having sincere cooperation based on the same concept of sustainability. This should include, especially when requested by the authorities concerned, coordination of action to improve the governance for river management. In the case of international donors, they should provide not only financial assistance to the national governments but also assist the governments to improve the governance system at each level of organization.

2.3. Qualification of These Links

Based on the historical background, the Mekong River Commission is a key regional body, which binds the four Lower Mekong River Basin countries in the context of river basin management. The planning tool used by the MRC is the Basin Development Plan (BDP), which is defined as the 'general planning tool and process that the Joint Committee would use as a blueprint to identify, categorize and prioritize the projects and programmes to seek assistance for and to implement the plan at the basin level.'[48] However, the BDP is only at the framework preparation stage. The plan has been criticized as programme approach including hydrology; environment; remote sensing and GIS; subbasin and project planning, and the various sectors of forestry, agriculture, irrigation, fisheries, and transport. The projects are predominantly in the transport and energy sectors. The MRC as an international organization, is able to receive funding from various sources including foreign aid in order to carry out the projects.

The MRC Secretariat is tasked to define particular projects but it is not responsible for their implementation. Therefore, the MRC lacks the power to deal with the final outcomes of the projects or even the stages beyond the early investigative phase of the projects. This is the weakness of the organization. Ideally, it needs to restructure and reform the power of authorization and even enforcement and regulation of the natural resource use for the whole Mekong River Basin, not only for the Lower Basin.

Economic development is a major goal in all riparian countries' policy with national economic policies striving for high economic growth rate based on macroeconomic strategies. Consequently, the high competition among the riparian countries is aggressive and leads to resource competition regarding dependency and interdependency on natural resources in the Mekong River Basin (Figure 4). There is direct resource competition, particularly water supply for agricultural production, e.g., for rice, between Thailand and Vietnam, which are competitors in the

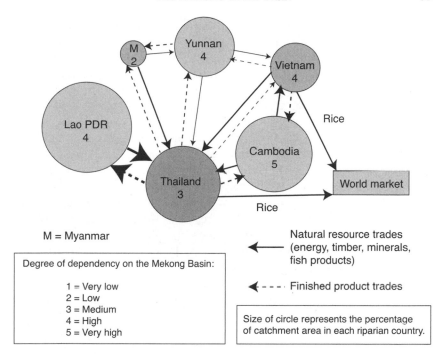

FIGURE 4. Interdependency on trade in natural resources between riparian countries in the Mekong River Basin

world rice market. The conflict on the diversion of water from the
Mekong River into the tributaries in the northeast of Thailand may
affect the quantity and quality of water supply throughout the agricultural land in the delta as well as in Vietnam.

There is indirect resource competition for energy production,
with a high potential for hydropower production though the riparian
countries that seem to have high energy demands for their own countries especially Thailand. Thailand has limited options for energy
production and it is necessary to import from neighboring countries.
Riparian countries that have surplus potential of energy production
can become competitors in the energy market. For instance, the economies of Myanmar, Lao PDR, and Cambodia are growing fast and
depend in large part on neighboring countries by exporting natural
resources such as timber, fishery products, and mineral products, with
electricity in particular to Thailand. In total, trade and markets still
underlie commodification of the resource base and drive the region.
Foreign direct investment has been a major force for economic growth

with Thailand by notably increasing investment in neighboring countries. The riparian countries are attracting various sources of financial funding such as Japan, Taiwan, Hong Kong, Singapore, and South Korea.[49] The development assistance to Lao PDR, Cambodia, and Vietnam has been provided from the European Union (EU), Australia, Scandinavia, and UN agencies as well as from World Bank and the IMF.

Significant conflicts underlie many of the necessary initiatives in the Mekong Basin natural resource management. It needs some lateral thinking in aid delivery that aims to help reduce conflict to a manageable level. Within the MRC framework, areas of conflict other than straightforward country-to-country interests are those requiring most attention, analysis, and action. For instance, the problems of resource development are likely to disadvantage particular areas, sections of the population, or social groups in order to benefit others and there are inadequate compensatory or mitigating mechanisms. Therefore, it is necessary to clarify rights and responsibilities, particularly through progressive advances in defining tenure arrangements. Tenure must be equitable and promote more long-term stewardship of land, water, and forest resources. It is difficult to separate water resource management from land and forest management because they are strongly interlinked systems. Therefore a holistic approach is essential. Such improvements can provide an improved quality of life, livelihoods, and economic development. Therefore, the concept of stakeholder participation is essential to conflict resolution by providing fairness and sustainable resource use to all stakeholders in the region.

3. MANAGEMENT OF THE MEKONG RIVER BASIN

3.1. Assessment of the Actual Management of the Mekong River Basin

Political stability is essential to water resource management and to developing cooperation among the six countries. Cooperation had been delayed by the war in Indochina and by the partial dependence on the relative proportion of territory, population, and production at a regional scale. At the local scale, the limited access due to political insurgency, lack of infrastructure, poor governance among all stakeholders, and limited political representation have caused slow economic growth despite rapid natural resource utilization. Water resource management in each country in the Mekong River Basin is summarized as follows:

Yunnan (China): Due to its geographic position at the Upper Mekong Basin and to its political isolation, Yunnan has not been in contact with other riparian countries. The Mekong River is important to the people of Yunnan Province and to residents of the basin because the river remains a lifeline for people from Yunnan to Thailand via Lao and Myanmar. Yunnan provincial government is interested in engaging with the Lower Mekong countries with the aim of coordinating and developing areas of infrastructure investments, navigation with 'free navigation' agreement, and ecotourism with the Tourism Authority of Thailand.[50] In addition, the utilization of hydropower has received funding from foreign aid programs and from lending agencies for dam constructions. The dams on the Upper Mekong River (Table 7) will impact on the ecological characteristics, flow rate, flooding, and silting downstream, and furthermore will complicate any improvement of the shipping system. Two of these dams are large storage dams, which primarily produce hydropower to meet Chinese needs and also for sale to neighbors, especially Vietnam and Thailand. There is considerable concern within Cambodia (where Tonle Sap is located) and from others in the regional and international community about the impact of the dams on the riverine ecosystems-fish ecology, sediment/nutrients transfer, and annual riverbank gardening. In relation to water governance, it epitomizes nontranparency, nonprovision of information, and noninvolvement of the public in learning about and influencing the associated decision-making process. From a natural resource management perspective, Yunnan provincial authority is interested in the utilization of water resources more than the improvement of good governance among the riparian countries. For instance, China intends to cover only a few aspects of water resource management (e.g., flood control, some pollution control, and incomplete efforts at water sharing) and do little in the way of full natural resource assessment and monitoring and basin wide policy making.

Myanmar: Myanmar is still subservient to the political, military, and economic policies following the political and military conflicts. Myanmar is still not a member of the MRC and the government is showing little concern for resource management and development as it is a relatively minor player in the basin. The State Law and Order Restoration Council (SLORC) does not have full control of the areas of Myanmar within the basin and within neighboring Kayah State and no resource management projects are planned. The only major concerns relate to forest management because of the importance of timber exports, mainly to China and Thailand, particularly for military purposes. With the strong relationships between China and Myanmar, it is likely that Myanmar does not have any negative reactions to the upper-stream development. Water resource management is not a recognized priority for domestic politic authorities and for the central government.

Lao PDR: Since the end of the cold war and increasing political rapprochement, the opportunities for development of the Mekong Basin have been greatly enhanced through access to foreign aid institutions or loan agencies. With the abundant watershed and steep topography, Lao PDR has the largest hydropower potential and the most significant resource among the riparian countries. These attract foreign investment for economic development. Therefore, Lao PDR is in competition for the hydropower market with China, Myanmar, and Vietnam, with plans for large hydropower developments to export energy to Thailand. Lao PDR authorities may be aware of the negative impacts on deforestation, greater sediment loads, and floods from the mainstream development projects. On the other hand, they are not concerned with the vital impacts on the tributary development. These hydropower development projects are important for earning foreign exchange and for improving the economic performance. However, they pose numerous risks to re-source sustainability. Due to the lack of internal sources, the Lao PDR government needs to rely on the advice of numerous foreign consultants and agencies in resource management. The Science, Technology, and Environment Agency (STEA) was founded by the prime minister in 1993, and serves as the center for the coordination of science, technology, and environment-related activities. The STEA is the national focal point for environmental management legislation and for assess environmental impact reports. However, the organization is still understaffed and underskilled, and human resource development needs to be given high priority in order to deal with natural resource management. Thus, natural resource management in Lao PDR is still at the early stage concerning development policies and action plans.

Thailand: About one-third of its territory and population is located within the Mekong Basin. It is less dependent on the Mekong Basin and on its resources than Lao PDR and Cambodia.[51] Over the past two decades, Thailand's rapid development plus the rapid economic growth rate in the region, have resulted in the rapid loss of forest cover and increased environmental pollution. Since the 1980s, environmental degradation and mounting public pressure have forced the Thai government to incorporate the environment into the national policy agenda. Thailand's approach to natural resource management in the Mekong River Basin is characterized by an increasing capacity to secure positive environmental outcomes through the National Environment Board (NEB) and through other government agencies interacting with local people, academia, and NGOs. Since the promulgation of the National Environmental Quality Act (NEQA) in 1992, it mostly promoted environmental management plans and policies both short-term and medium-term. Unfortunately, government entities either at the center or in the provinces treat environmental plans mechanistically as a budgetary framework, while the problems of natural resource depletion resulting from people's unchanged attitude and behavior are not addressed.[52] Moreover, there are still major obstacles to achieving sustainable resource management practices because of the weak coordination between government agencies, the centralized administration system focused on Bangkok, and the complex environmental problems resulting from previous development projects. There is an important reason for this. The State's management of the environment has been marred by an undue concentration of power in the hands of government agencies, leaving the private sector and in particular, the localities directly affected by the state's decision making to play second fiddle, and resulting in injustice, lack of transparency, and discrimination.[53]

Cambodia: Its resource management is significantly concerned with the implications of river development on the Tonle Sap system, which is the heart of the country's agricultural and fisheries production. With 50 percent of Cambodia's GDP coming from agriculture, fisheries, and forestry, the Mekong Basin is an important natural resource unit for the people of Cambodia. Political, economic, and security uncertainties, plus difficulties in implementing an integrated approach of natural resource management within central government agencies or between different levels of government, have induced rapid natural resource exploitation with the growing economy. Since 1993, both the Ministry of Environment and the Ministry of Rural Development have been established, but they have very limited influence and authority in matters concerning resource management. Instead, the Ministry of Agriculture still controls decisions on resource use and management. The approach to natural resource management in Cambodia is still underdeveloped in terms of empowering legislation for the Ministry of Environment to have full authority within the decision-making process. Other weaknesses are the poor institutional capacity, coordination and support, the lack of skilled staff, dependence on aid flow, and poverty. These reflect the need for external aids and consultations in order to solve the problems of environmental degradation, deforestation, and subsequence siltation. To strengthen institutional capacity, the Mekong River Commission head office has been moved from Bangkok, Thailand, to Panompeh since 1999, which provides skilled staff for building capacity for sustainable resource management in terms of policy and action plans.

Vietnam: Twenty-nine percent of the territory is in the basin, represented by the Central Highlands and the Mekong Delta. The Central

Highlands are important for upland agriculture and hydropower development. The Mekong Delta produces 50% of the rice crop and 40% of the total agricultural output. Increased agricultural production has been the strategy of economic renovation since the mid-1980s, but natural resource management has been a lower priority than economic development. Within Vietnam, the different approaches to water management are due in part to the different geographic characteristics of the country and associated socioeconomic activities. This imperative for economic growth is only slowly beginning to be regulated by broader concerns for the environment. Recently, the Vietnam government has established institutions, legislations, policies, and processes for environmental protection and natural resource management. The National Environment Agency (NEA) was established by the Ministry of Science, Technology and Environment (MOSTE) in 1993, to strengthen environmental protection and management. In addition, the National Research Programme on Environment (NRPE) was reformed in 1991, to provide a formal process for academic input to policy formulation through research such as environmental impact assessments. Until 1999, Vietnam adopted a "Law on Water Resources" that states that organizations and individuals have the right to exploit and use water resources. At the same time, they have a responsibility to protect water resources. The principles of "user pays" and "polluter pays" are included.[54]

The MRC is the regional body concerned with resource management development policies and action plans among the riparian countries. The overall objectives are to accelerate sustainable development and growth within the basin, in accordance with the Basin Development Plan and to form close partnerships with constituencies among the donor communities (MRC, 1995a). J. Dore indicated that the Mekong River Basin water utilization negotiations are another key regional environmental governance example.[55] Negotiations commenced in earnest during 2000, facilitated by the Phnom Penh-based MRC Secretariat (MRCS). They are conducted as part of the Water Utilisation Program (WUP) process (1999–2005). The WUP is 'legitimized' by a signed intergovernment agreement—the 1995 Mekong River Agreement that requires water use to be negotiated between Thailand, Vietnam, Cambodia, and Laos as indicated in the following articles:

- Article 1 commits the signatories to cooperate in all fields of sustainable development, utilization, management, and conservation of the water and water-related resource of the Mekong River Basin, including, but not limited to, irrigation, hydropower, navigation, flood control, fisheries, timber floating, recreation, and tourism.

- Article 3 pledges signatories to protect the environment, natural resources, aquatic life and conditions, and ecological balance of the Mekong River Basin from pollution or from other harmful effects resulting from any development plans and uses of water and water-related resources.
- Article 5 commits members to reasonable and equitable utilization.
- Article 6 commits members to maintain mainstream flows.
- The challenge for the WUP is to put Articles 5 and 6 into practice (Art. 26 requires the MRC to prepare water utilization rules to enforce Arts. 5 and 6).

Therefore, the MRC is now in a position to deliver the 'best practices' in Integrated River Basin Management (IRBM). It has strong legislation that ensures consultations and evaluation of major development projects. It also has a strong strategic focus and has recently adopted an EIA process that must be followed for all major development projects. It is developing a strong information and modeling base and is presently working up an appropriate approach to community awareness and participation.[55]

The next section considers how assessment of the impacts of hydropower development could be revised to improve the balance between social, economic, and environmental information in the decision-making process—an essential basis for comprehensive governance.

3.2. How the Situation is Likely to Evolve

Earlier parts of this book have illustrated the influence of global economic and environmental policies. The differing national circumstances and priorities in the Mekong River Basin have highlighted the importance of hydropower development with many proposed projects (Table 7). But how do these influences and interests interact in specific circumstances and at the local level? How can the social, economic, and environmental impacts be assessed, compared, and integrated into Impact Assessment (IA) across all scales? Conventional IA emphasizes the immediate economic costs and benefits but pays little attention to qualitative, long-term, regional, and global impacts. How can the system be improved and decision-makers provided with greater opportunity to assess the various options?

Recent reviews of the IA process for development projects indicate a number of weakness,[57] including

- Inadequate determination of the spatial context of the project.
- Poor or insufficient baseline information and treatment of biodiversity such as a simple 'list' of species found in a project area.

- Lack of rigor in the cost-benefit analysis.
- Insufficient attention to implementation and monitoring of mitigation measures and environment management plans including institutional arrangements.

IA is a critical potential tool for regulating environmental practices of both national and multi-national organizations reflecting the holistic concept of sustainable resource management.[58] To improve the IA performance it is necessary to systematically assess, quantify, and evaluate the ecological, social, and economic consequences as part of the cost-benefit analysis of projects.[59] Traditionally, biodiversity has not been addressed within the IA[60] until after the implementation of institutional agreements among international organizations (CBD, Ramsar, UNDP) have been accepted. Assessment of biodiversity has been constrained by difficulties in monetary evaluation in relation to national, regional, and local economies. While monetary evaluation of biodiversity can be of some use, it may not reflect the inherent or long-term ecological value. Furthermore, conventional IA usually considers the local cost and benefits, with little consideration for regional and global effects.

Therefore, an alternative impact assessment approach was recommended and adopted aiming to explore options for the assessment of

- Qualitative and quantitative measures of biodiversity, including ecosystem services and human elements.
- Nonmonetary as well as monetary evaluation of biodiversity.
- Consideration of development consequences at local, regional, and global scales.
- Integration of different measures or indexes into decision support models.

The aim is to provide an improved approach to an assessment of management options, incorporating biodiversity evaluation, exploring the resources utilization with shared benefits to environment, social, and economic sectors, and minimizing negative impacts from hydropower development projects. Although the focus is not only on dam construction projects but also on other types of development projects, the new alternative or integrative impact assessment approach should apply and provide more information for the decision-making process.

4. FURTHER CONSIDERATIONS

From the wider review of the Mekong River, a number of other more general actions should be considered, at global, regional, and local levels. The concept of comprehensive water governance is similar to an "integrating" plus "sustainability" concept that is the most difficult to both understand and implement across different dimensions or levels (Figure 5). The success of comprehensive water governance can be implemented at, or it might have feedback loops from, any level. However, the implementation must restructure policies and actions with regard to the holistic approach, sustainable resource use, and equitable sharing of benefits between environmental, social, and economic sectors.

Global communities:
- Promote management for carbon storage as one of a number of ecosystem services and as a simple, rational, and applicable indicator for biodiversity at an ecosystem level.
- Link biodiversity and ecosystem services, including carbon storage, to achieve more equitable sharing of natural resource benefits.
- Commit between developed and developing countries to improve utilization of natural resources based on concepts of sustainability over extended time scales, with promotion of principles of comprehensive governance.
- Establish institutions responsible for fair trade of carbon between developing and developed countries aiming to mitigate increases in atmospheric carbon and to reduce fossil fuel consumption.

Regional and national levels:
- Harmonize the various sectors in environmental, social, and economic policies and plans into the national operation.

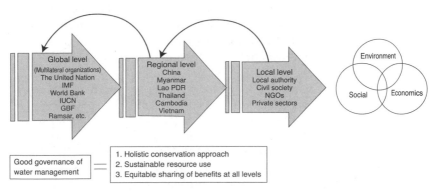

FIGURE 5. A concept of comprehensive water governance in the Mekong River Basin

- Review environmental law and institutional mechanisms to strengthen links between management for biodiversity conservation and carbon sequestration.
- Ensure socially responsible national economic development while protecting biodiversity for future generations.
- Establish environmental policy with zero discount rate of biodiversity value, especially carbon storage value, to reduce discrimination in favor of projects with low capital and high operating costs.
- Promote strategy of carbon storage benefits, including tradable carbon emission permits system in the international market related to the Clean Development Mechanism (CDM) from the Kyoto Protocol.
- Strengthen agreement between countries to accept liability for the performance of the industrial countries in clean energy options.

Local level:

- Engage stakeholders in local government, community, and private sectors in development planning from an early stage.
- Provide improved environmental education to all stakeholders in the communities.
- Encourage commitment of stakeholders to sustainable resource use as a means of improving living standards.
- Encourage shared responsibility among stakeholders across sectoral activities and provide effective bargaining power with the government regarding development projects or policies.
- Create a sense of environmental awareness and efficient utilization of biodiversity within the communities.

Articulation of a collaborative ethos based on the principles of consultation, transparency, and accountability is a fundamental requirement. Governments bare the responsibility for constructive responses to the issue of water management affecting their peoples, other peoples in the basin, and the global community. Adoption of a more comprehensive IA approach within national policy would be an important step in water resource governance, encouraged by multi-lateral and international organizations. At the local level, direct implementation will only be possible through enhanced education, but any adequate system of comprehensive governance must have the capacity to control and deploy the water resources. The system must encompass stakeholders with the responsibility and vision to achieve results, combined with the necessary safeguards to protect and sustain social, economic, and environmental resources. Finally, it will require a transition from arbitrary power to agreed comprehensive governance within the Mekong River Basin.

NOTES

1. Governance is the exercise of political, economic, and administrative authority in the management of a country's affair at all levels. Governance is a neutral concept comprising the complex mechanisms, processes, relationships, and institutions through which citizens and groups articulate their interests, exercise their rights and obligations, and mediate their differences (UNDP, 1997).

2. MRC, 1996.

3. Ibid.

4. MRC, 1994.

5. MRC, 1996.

6. The Himalayas zoogeographic zone is defined as an Indo-Chinese subregion of the Oriental zoogeographic region; the distribution of fauna from the Indo-Chinese subregion has been prevented from spreading northward by the Himalayas (Beaufort, 1951 and Lincoln, Boxshall, and Clark, 1998).

7. The Sunda zoogeographic zone or the Greater Sunda Islands is defined as the southern oriental region; the distribution of fauna from Sumatra, Borneo, and Java has been merged northward through the Malayan peninsula (Beaufort, 1951).

8. MRC, 1996.

9. *Melaleuca* forests predominate on acid-sulphate soils. These forests provide *Melaleuca leucadendron* timber for construction material (valuable due to its strength, durability, and resistance to insect and fungi attacks) (MRC, 1996).

10. Hirsch and Cheong, 1996.

11. MFNN, 1996a.

12. MFNN, 1996b.

13. MRC, 1996.

14. Hirsch and Cheong, 1996.

15. Ibid.

16. MRC, 1996.

17. Ibid.

18. Ibid.

19. Ibid.

20. Gajaseni, 2000.

21. Chapman and Daming, 1996.

22. MRC, 1996.

23. Grainger, 1993.

24. MRC, 1996.

25. IUCN, 1996.

26. MRC, 1996.

27. Maltby, 1999.

28. FAO, 1999.

29. Gajaseni, 2000.

30. Dore, 2001.

31. World Bank, 1999.

32. MRC, 1994.

33. Berman, 1998.

34. MRC, 1994.

35. Ibid.
36. Tu, 1996.
37. Hempel, 1996.
38. UNDP, 1998.
39. Hirsch and Cheong, 1996.
40. MRC, 1996.
41. WCD, 1999.
42. Hirsch and Cheong, 1996 and Dore, 2001.
43. Hirsch and Cheong, 1996.
44. UNDP, 1997.
45. Ibid.
46. IMF, 1997.
47. Ibid.
48. MRC, 1995.
49. Hirsch and Chong, 1996.
50. Berman, 1998 and Hilton, 1998.
51. Pednekar, 1998.
52. Chomchai, 2001.
53. Ibid.
54. Can, Phan, and An, 2001.
55. Dore, 2001.
56. WCD, 1999.
57. World Bank, 1997 and Sadler and Verheem, 1996.
58. Hempel, 1996.
59. World Bank, 1997 and IAIA, 1999.
60. World Bank, 1997.

BIBLIOGRAPHY

Beaufort, L. F. de (1951). *Zoogeography of the Land and Inland Waters.* Sidgwick and Jackson: London: 60–90.

Berman, M. L. (1998). *Opening the Lancang (Mekong) River in Yunnan: Problems and Prospects for Xishuangbanna.* Master's thesis, University of Massachusetts.

Can, Le Thac, Do Hong Phan, and Le Quy An (2001). Environmental Governance in Vietnam in Regional Context. In *Mekong Regional Environmental Governance: Perspectives on Opportunities and Challenges 2001.* The Resources Policy Support Initiative (REPSI), Chiang Mai, Thailand: 13–26.

Chapman, E. C. and H. Daming (1996). *Downstream Implications of China's Dams on the Lancang Jiang (Upper Mekong) and Their Potential Significance for Greater Regional Cooperation, Basin wide.* <http://asia.anu.edu.au mekong dams.html>.

Chomchai, P. (2001). Environmental Governance: A Thai Perspective. In *Mekong Regional Environmental Governance: Perspectives on Opportunities and Challenges 2001*. The Resources Policy Support Initiative (REPSI). Chiang Mai, Thailand: 117–137.

Dore, J. (2001). "Environmental Governance in the Greater Mekong Sub-region." In *Mekong Regional Environmental Governance: Perspectives on Opportunities and Challenges 2001*, The Resources Policy Support Initiative (REPSI). Chiang Mai, Thailand: 191–250.

FAO (1999). *State of the World's Forest 1999*. FAO, Rome.

Gajaseni, N. (2000). *An Alternative Approach to Biodiversity Evaluation: Caste Study in the Lower Mekong Basin*. Ph.D.—diss., University of Edinburgh.

Grainger, A. (1993). *Controlling Tropical Deforestation*. Earthscan Publications, London.

Hempel, L. C. (1996). *Environmental Governance*. Island Press, Washington DC.

Hilton, P. (1998). *Resource Management in the Yunnan Mekong Basin*. Asia Research Centre. Australia: Murdoch University.

Hirsch, P. and G. Cheong (1996). *Natural Resource Management in the Mekong? River Basin: Perspective for Australian Development Cooperation*. University of Sydney.

International Association for Impact Assessment (IAIA) (1999). Impact Assessment and the Biodiversity Agenda. In *The Proceedings of IAIA Conference 1999*. Glasgow, Scotland.

International Monetary Fund (IMF) (1997). *Good Governance*. International Monetary Fund, Publication Services, Washington DC.

International Rivers Network (IRN) (1999). *Power Struggle: The Impacts of Hydro-Development in Laos*. International Rivers Network.

IUCN (1996). The 1996 IUCN Red List of Threatened Animals. IUCN, United Kingdom. http://www.wcmc.org.uk/cgi-bin/arl-output.p.

Lincoln, R., G. Boxshall, and P. Clark (1998). *A Dictionary of Ecology, Evolution and Systematics*. Cambridge University Press, Cambridge.

Maltby, E. (1999). Wetland Ecosystem Functioning: An Expert System-type Approach to Support Decision-making. In *Ecosystem Management: Question for Science and Society*, E. Maltby, M. Holdgate, M. Acreman, and A. Weir (Eds.). College Hill Press, United Kingdom: 131–140.

Mekong Fisheries Network Newsletter (MFNN) (1996a). "1,200 Different Fish in the Mekong Basin." *Mekong Fish Catch and Culture* 2(1): 4–5.

Mekong Fisheries Network Newsletter (MFNN) (1996b). "1,000,000 tonnes of fish from the Mekong." *Mekong Fish Catch and Culture* 2(1): 1.

Mekong River Commission (MRC) (1994). *Mekong Mainstream Run-of-River Hydropower: Main Report.* Bangkok.

Mekong River Commission (MRC) (1995a). *Mekong River Commission towards Sustainable Development: Annual Report 1995.* The Mekong River Commission Secretariat, Bangkok.

Mekong River Commission (MRC) (1995b). *Pre-feasibility Study on Ban Koum Run-of-River Hydropower Project (Basinwide).* The Mekong River Commission Secretariat, Bangkok.

Mekong River Commission (MRC) (1996). *Mekong River Basin Diagnosis Study.* Bangkok.

Nilsson, S. and W. Schopfhauser (1995). The Carbon-Sequestration Potential of a Global Afforestation Program. *Climatic Change* 30(3): 267–293.

Pednekar, S. (1998). *Resource Management in the Thai Mekong Basin.* Asia Research Centre, Australia: Murdoch University.

Sadler, B. and R. Verheem (1996). *Status, Challenges and Future Directions.* Ministry of Housing, Spatial Planning and the Environment, The Netherlands.

Thailand Environment Institute (TEI) (1998). *Asia Least-Cost Greenhouse Gas Abatement Strategy: Thailand Third Draft Final Report Submitted to Office of Environmental Policy and Planning, The Government of Thailand.* Thailand Environment Institute, Bangkok.

Tu, T. D. (1996). Sustainable Development in the Mekong River Basin. *Land Lines* 8(3): 1–3.

United Nations Development Programme (UNDP) (1997). *Governance for Sustainable Human Development: A UNDP Policy Document.* United Nations Development Programme, New York.

United Nations Development Programme (UNDP) (1998). *Fourth Quarter Report of UNDP Thailand.* http://www.undp.or.th/news-reports/4quarep.asp.

World Bank (1999). Country Data: Myanmar, Lao PDR, Thailand, Cambodia and Vietnam. http://www.worldbank.org/data/countrydata.html.

World Commission on Dams (WCD) (1999). *River Basin Management—Its Role in Major Water Infrastructure Project: River Basins—Institutional Framework Management Option.* http://www.oieau.fr/riob/forum2/rom_draft_151199.htm.

World Resource Institute (WRI) (1999). Facts and Figures: Country Environmental Data (Asia). http://www.wri.org/facts/cs-asia.html.

Chapter 3

The Danube River Basin

Stephen McCaffrey

1. GENERAL CONTEXT

1.1. The Biophysical Dimension

The Danube is the second longest river in Europe (after the Volga), flowing approximately 2,857 km from its source in the Black Forest region of Germany, only 40 km from the Rhine, to its mouths in the Black Sea in Romania and Ukraine. Especially since the completion of the 171-km-long Rhine-Main-Danube canal in September 1992,[1] it is at least potentially one of the principal transportation arteries on the European continent, linking the landlocked states of Austria, Slovakia, and Hungary with the North and Black seas. The operation of at least certain of the numerous dams on the river (chiefly the Iron Gates dam), together with navigational hazards (principally siltation in some sections and seasonal flooding and ice) have prevented the Danube from reaching its potential as a navigational route. The Danube is the only major European river to flow from west to east.

The Danube Basin covers 817,000 sq km² in seventeen countries in the heart of central Europe. It flows through all of Hungary; most parts of Romania, Austria, Slovenia, Croatia, and Slovakia; and significant parts of Bulgaria, Germany, the Czech Republic, Moldova, and Ukraine. Portions of the territories of the Federal Republic of Yugoslavia (Serbia and Montenegro), Bosnia, and Herzegovina and small part of Italy, Switzerland, Albania, and Poland are also included in the basin. The Danube's average runoff is 0.206 km³/year. A lengthy drought during the past decade has reduced flows and water availability for consumptive uses in Lower Basin States. However, severe flooding during the summer of 2002 caused considerable damage in the basin.[3]

The Danube Basin may be divided into three parts, or subbasins, which are separated from each other by higher landforms, or "gates": The Upper Danube subbasin lies in Germany and Austria. The "Devin Gate," on the boundary between Austria and Slovakia at the confluence of the Morava River, marks the beginning of the Middle Danube region, which includes portions of Slovakia, Hungary, Croatia, and Serbia-Montenegro. While flowing along the border between Yugoslavia and Romania the Danube passes through the "Iron Gate," entering the Lower Danube subbasin, which comprises portions of Romania, Yugoslavia, Bulgaria, Moldova, and Ukraine. While a navigation canal between the Romanian cities of Cernavoda and Agigea connects the Danube with the Black Sea, its natural bed has already begun turning north at the former city, then bends eastward at Galati, near Moldova. The river then divides into three distributaries in the Danube Delta, the northernmost forming the boundary between Ukraine and Romania, the middle emptying into the Black Sea at Sulina, usually considered the mouth of the Danube, and the southern ending at the Romanian city of Sfintu Gheorghe.

The Danube is navigable by ocean vessels to Braila, Romania, and by river craft as far as Ulm in Germany, a distance of 2,600 km. However, the Rhine-Main-Danube canal permits navigation by river-going vessels from the North to the Black seas. Some 60 of the approximately 300 tributaries of the Danube are navigable. The principal ones, in order in which they merge with the Danube, include the Lech, Isar, Inn, Morava, Váh, Raab (Raba), Drava (Drau), Tisza, Sava, Siret, and Prut. Winter ice and seasonally varying water levels, especially flooding after spring thaws, may impede navigation.

A large number of dams—including hydroelectric plants—dikes, navigation locks, and other hydraulic structures have been built, chiefly in the Upper Basin States, to serve the river's important function. According to one study there are 49 planned or existing hydropower stations on the Danube, 40 of which are located in Germany and Austria.[4] These structures form over forty reservoirs on the Danube River itself. Important cities on the river include Ulm, Regensburg, and Passau, in Germany; Linz and Vienna, in Austria; Bratislava, in Slovakia; Budapest, in Hungary; Belgrade, in Serbia; and Galati and Braila, in Romania. Canals link the Danube to the Main, Rhine, and Odra (Oder) rivers. The Danube Basin supports the supply of drinking water, agriculture, industry, fishing, tourism and recreation, power generation, navigation, and the end disposal of wastewaters of the countries through which it flows. However, the principal uses of the Danube are navigation and hydroelectric power generation. These uses have been the subject of a number of negotiations between riparian States.

The States of the Upper Basin, Germany, and Austria, enjoy a higher level of development than their downstream co-riparians and rely on the river principally for power generation, industrial uses, and waste disposal. The less-developed countries in the Lower Basin look to the Danube more for drinking water, irrigation, and fisheries.

The Danube River Basin constitutes an aquatic ecosystem that is internationally recognized for its richness. The wide variety of habitats in the basin support significant species and genetic diversity. The delta of the Danube River is the second largest natural wetland area in Europe, a region of marshes and swamps, broken by tree-covered elevations. The delta is home to a wide variety of species of plants, fish, birds, and mammals, some of which are endangered.[5] The waters of the Danube River Basin and its tributaries combine to make up an aquatic ecosystem of high economic, social, and environmental value. The basin contains numerous important natural areas including wetlands and inland floodplain forests.

The Danube has always been an important transportation and commercial route between Western Europe and the Black Sea. However, intensive agricultural, industrial, and urban uses have created problems of water quality and quantity and reduced biodiversity, threatening the health of the basin's ecosystems. The massive structures built on the river have caused changes in flow patterns and have affected the functioning and biodiversity of the river system. All of these influences have caused significant environmental changes, such as reduced sediment transport, increased erosion, and reduced self-purification capacity. Anthropogenic change has also affected its use for drinking water supply, recreation, and bathing.

1.2. The Historical and Legal Context

The Danube River (ancient Danuvius) formed, in the third century AD, the northern boundary of the Roman Empire in southeastern Europe. Early in the Middle Ages migratory peoples crossed the Danube on their way to invade the Roman, and later the Byzantine, empires. In the nineteenth century it became an essential link between the growing industrial centers of Germany and the agricultural areas of the Balkan peninsula.

The river has a fascinating political and legal history, which has been recounted elsewhere.[6] While the Central Commission of the Navigation of the Rhine has been described as the doyen of international organizations,[7] countries in the Danube Basin, as well as nonriparians, have cooperated through institutions in the field of navigation since the Treaty of Paris in 1856.[8] Nonnavigational uses were not addressed significantly in Danube Basin treaty law until the end of the First World

War. The 1920 Treaty of Trianon[9] established a Hydraulic System Commission that regulated the nonnavigational uses of much of the Danube Basin,[10] foreshadowing the modern drainage basin approach to the management of international watercourses. The treaty also provided that in the resolution of disputes due allowance would be made for all rights relating to irrigation, waterpower, fisheries, and other national interests.[11] These could, in fact, be given priority over navigation with the consent of the States concerned.

The 1919 peace treaties of Versailles and St. Germain-en-Laye internationalized most major European rivers, including the Danube. They provided the basis for the Paris Convention of 1921 that established an International Commission with jurisdiction over the river as far downstream as Braila, Romania, and left the European Commission of the Danube with authority over the remainder, from Braila to the Black Sea. After the Second World War, Danube riparians of Eastern Europe concluded the Belgrade Convention concerning the Regime of Navigation on the Danube of August 18, 1948,[12] which governs navigation on the river to this day. Western governments have not become parties to the Belgrade Convention, but Germany and Austria have been participating in the work of the Danube Commission since 1960. It has been commented that this organization is a "mere organ of coordination" and that the "real power is vested in the riparian states, which exercise full control over the stretches of the river within their territories [with regard to navigation] with the proviso that they must respect freedom of navigation."[13]

But by the late 1970s it had become clear that nonnavigational uses and other issues relating to the Danube, particularly water quality, had taken on more importance.[14] In 1985 eight of the riparian countries, meeting at Bucharest, signed the nonbinding Declaration of the Danube Countries to Cooperate on Questions Concerning the Water Management of the Danube, or the "Bucharest Declaration."[15] Then in September 1991, Danube Basin countries met with interested governments and international institutions in Sofia to draw up an initiative to support and reinforce national actions for the restoration and protection of the Danube River—the Environmental Programme for the Danube River Basin (known as the Danube Environmental Programme). A Task Force and Programme Coordination Unit were established in this connection. The Danube Environmental Programme supported monitoring, data collection and assessment, emergency response systems, and reinvestment activities that provided for analysis of seventeen tributary catchments in the basin. Alongside these, the programme supported institutional strengthening, capacity building, and NGO activities.

This effort led the Danube River Basin countries and the European Union to sign the Convention on Cooperation for the Protection and Sustainable Use of the River Danube (the Danube Protection Convention) on June 29, 1994 in Sofia.[16] The convention, which entered into force on 22 October 1998 and furthers the aims of the 1992 ECE Helsinki Convention on the Protection and Use of Transboundary Watercourses and International Lakes and the Convention on the Protection of the Black Sea against Pollution of the same year, has as its general objective the achievement of "sustainable and equitable water management" (Art. 2.1). Other principal objectives of the convention are "to control the hazards originating from accidents involving substances hazardous to water, floods and ice hazards" and to reduce pollution of the Black Sea from land-based sources within the basin (Art. 2.1). The parties agree to "at least maintain and improve the current environment and water quality conditions of the Danube River and of the waters in its catchment area and to prevent and reduce as far as possible adverse impacts and changes occurring or likely to be caused." (Art. 2.2.). The parties recognize the "urgency of water pollution abatement measures" (Art. 2.3) in order to ensure the "conservation and restoration of ecosystems" and to satisfy public health requirements (Art. 2.3). The Convention provides that the polluter pays and precautionary principles are "a basis for all measures aiming at the protection of the Danube River and of the waters within its catchment area" (Art. 2.4.). It also stipulates that work being performed by the Contracting Parties under the 1985 Bucharest Declaration is transferred to the framework of the Convention (Art. 19).

Significantly, the parties to the Danube Protection Convention agreed to establish an International Commission for the Protection of the Danube River (Art. 18 and Annex IV) as a legal entity with a Permanent Secretariat, headquartered in Vienna (Annex IV, Arts. 7 and 10). The general mission of the commission is to assist the parties in implementing the objectives and provisions of the convention and to provide a framework for cooperation between them. The commission is charged with elaborating proposals and recommendations to the parties, receiving reports submitted by the parties pursuant to Article 10 of the convention, making proposals concerning amendments or additions to the convention on the basis of experience with its implementation and preparing draft regulations concerning the protection and management of the Danube and waters within its catchment (Art. 18, paras. 1, 4, and 5). Upon the entry into force of the convention, responsibility for the water-related parts of the Environmental Programme were transferred to the commission and to its secretariat.

A discussion of the legal context of the Danube would not be complete without at least brief mention of the Gabcikovo-Nagymaros Project case between Hungary and Slovakia.[17] The case involved a treaty concluded in 1977 by Hungary and Czechoslovakia providing for the construction of a major project consisting of a series of dams (also referred to in relevant documents as locks and barrages) and other works on a 200-km-stretch of the Danube River, between Bratislava and Budapest.[18] According to the treaty's preamble, the system of works is designed to attain "the broad utilization of the natural resources of the Bratislava-Budapest section of the Danube river for the development of water resources, energy, transport, agriculture and other sectors of the national economy of the Contracting Parties." As summarized by the International Court of Justice, the project "was thus essentially aimed at the production of hydroelectricity, the improvement of navigation on the relevant section of the Danube and the protection of the areas along the banks against flooding."[19] The treaty also provides for the protection of the quality of Danube waters and of the observation of obligations for the protection of nature. Project details are to be planned and implemented in accordance with a Joint Contractual Plan to be drawn up by agencies of the two States on the basis of the general framework set out in the treaty.[20]

As originally planned, the project consisted of a dam to be constructed by Hungary at the Hungarian town of Dunakiliti, which would create a reservoir and divert most of the flow of the Danube through a 31-km-bypass canal on Czechoslovak territory. A second dam, to be constructed by Czechoslovakia and equipped with shipping locks and a power plant with eight turbines, would be located on this canal at the Czechoslovak town of Gabcikovo. A third dam would be constructed by Hungary considerably farther downstream, after the Danube enters Hungarian territory, at the Hungarian town of Nagymaros. The dams at Gabcikovo and Nagymaros were to be operated in a coordinated manner, constituting a "single and indivisible operational system of works."[21]

Concerns relating to the effects of the project on the environment began to surface in the nascent civil society of Hungary in the 1980s. This criticism contributed to Hungary's suspension of work on the project in 1989, subsequent abandonment of work on the project as a whole, and an announcement in May 1992 that it was terminating the 1977 treaty.

Czechoslovakia objected strenuously to Hungary's withdrawal from the project and rejected Hungary's purported termination of the agreement as being without legal foundation. Since the project as originally designed could not be brought into operation without Hungary's participation, Czechoslovakia began investigating possible alternative solutions. It ultimately responded to Hungary's abandonment of work on the project and the May 1992 declaration of termination of the treaty through what it termed a *provisional solution*: damming the Danube in October 1992 at Cunovo, on a portion of the river entirely within its territory and upstream from Dunakiliti and Gabcikovo. This dam and related works, known as Variant C,[22] enabled Czechoslovakia to channel much of the flow of the Danube through the project's bypass canal and thus to put the upstream portion of the project into partial operation.

Slovakia became an independent state on January 1, 1993 and succeeded to Czechoslovakia's interest in the project by agreement with the Czech Republic. By Special Agreement of April 7, 1993,[23] Hungary and Slovakia submitted the dispute to the International Court of Justice (ICJ). Thus, while mediation efforts of the Commission of the European Communities did not result in the resolution of the dispute, they did eventually lead the parties to submit it to the ICJ.

The Court held in essence that treaty law did not allow Hungary to terminate the 1977 agreement; that Hungary was therefore not entitled to suspend and to subsequently abandon work on the project; that Czechoslovakia was not entitled to put into operation Variant C; and that the parties were obligated to negotiate in good faith "in light of the prevailing situation," and to "take all necessary measures to ensure the achievement of the objectives of the [1977 treaty]." The Court thus recognized that even though the treaty was still in force, the parties might decide to modify the project it provided for by, e.g., retaining Variant C and by eliminating the Nagymaros dam. The negotiations called for by the Court are still going on at this writing, more than five years after the Court rendered its judgment.

This dispute seems to have arisen in part because of forces unleashed by the waning power, and ultimately by the collapse, of the Soviet Union. Hungarian citizens, perhaps emboldened to speak out by the ripple effects of Perestroika in the Soviet Union, made their views known not only on the project but also on the government of Hungary itself, leading to the downfall of the old communist regime in 1989. The intense criticism of the project by elements of the Hungarian population appears to have made it difficult, as a matter of domestic politics, for Hungary to continue to observe the 1977 treaty—even if, as the Court found, the political transformation in the region did not amount to a "fundamental change of circumstances" under the law of treaties sufficient to release Hungary from its obligations under the 1977 agreement.[24] The fact that the negotiations mandated by the Court have, as yet, produced no agreement suggests that the project may still be unpopular in Hungary.

Yet the case highlights the importance of the Danube to the riparian states. As the Court observed, the river "has underlined and reinforced their interdependence, making international cooperation essential. . . . The cumulative effects on the river and on the environment of various human activities over the years have not all been favorable, particularly for the water regime. [¶] Only by international cooperation could action be taken to alleviate these problems."[25]

1.3. The Stakes

According to one study, the "most urgent issues" facing the Danube States are "differing standards for water quality among the co-basin countries; competing demands created by the exploitation of the river for the generation of electric power and transportation; and potential for the accidental releases of toxic chemicals."[26] Not surprisingly, it is the States enjoying a higher level of economic development in the Upper Basin that make most use of the river for industrial waste disposal, adversely affecting the States in the Middle and Lower basins that rely on the river for drinking water, irrigation, fishing, and tourism. However, economic development and resource-exploitation activities in the latter countries can have adverse impacts as well, as dramatically demonstrated by the cyanide spill into the Tisza River, the Danube's largest tributary, from the Aurul gold mine in Baia-Mare, Romania, on January 31, 2000.[27] The lack of adequate sewage treatment facilities in these countries poses serious problems as well. In the mid-1990s it could be said that "more than half of the wastewater in the former socialist states is untreated or receives only conventional primary treatment. The rest receives biological treatment, often, however, in Soviet-designed plants that are technically inferior and usually vastly overloaded."[28] The same study states that at that time, "water pollution from nutrients, oxygen-depleting substances, hazardous substances and microbiological contaminants is imposing risks to the region's ecology and the health of the people. The most important sources of this pollution include agricultural and livestock wastes and runoff, urban runoff, and industrial output from the chemical, pulp and paper, and mining and textile industries."[29]

The forty hydropower stations in Germany and Austria are equaled in energy output by the two colossal Iron Gates stations between the former Yugoslavia and Romania. "The Iron Gates hydroelectric power station was and remains controversial,"[30] in part due to severe bank and bed erosion problems downstream, especially in Bulgaria, and to the accumulation of heavy metals behind the 60-m-high dam. New hydropower or other major works have also become controversial, as demonstrated by the Gabcíkovo-Nagymaros project dispute between Hungary and Slovakia.[31]

River traffic, which could increase significantly as a result of the Rhine-Main-Danube canal, also poses significant threats. Especially worrisome is the risk of accidental spills of such hazardous substances as oil and chemicals, which would impact Lower Basin countries most severely.

The most urgent problems affecting the health of the Danube River ecosystems and the water users in the basin are the high nutrient loads (nitrogen and phosphorous), changes in river flow patterns and sediment transport regimes, contamination by hazardous substances including oils and cyanide, competition for available water, microbiological contamination, and contamination with oxygen-depleting substances.

But the interests of the Upper and Lower Basin countries differ considerably with regard to pollution control. While the Lower Basin countries would benefit greatly from improvements in pollution control, the Upper Basin countries, which have the resources to provide it, have the least to gain economically from improved water quality.[32] It is this dilemma that the Danube Environmental Programme and the Danube Protection Convention attempt to address, through a strategy involving integrated, participatory, and coordinated involvement of the riparian States.

2. ANALYSIS OF RELEVANT INSTITUTIONS AND ACTORS

The differences between the States of the Danube Basin, in terms of level of economic development, political/social culture, and environmental awareness, make coordination of action to protect Danubian ecosystems difficult and cooperation between riparians all the more important. One study has observed that "issues relating to the development and environmental protection of the river invoke a very different set of national and local actors in each country—different levels of responsibility, different administrative laws and procedures, different national priorities, and widely-differing resources for their solution."[33] But the collapse of the Soviet Union and the subsequent harmonization of the economic and political systems of countries in the basin, as well as the steady expansion of the European Union, have led to what observers have perceived as a change in attitude "from a view of exploiting the river for economic purposes to a view of integrated management of the river basin to promote sustainable development."[34]

The key actors for change are the public authorities—including international organizations such as the European Union—public and private enterprises, nongovernmental organizations (NGOs), and the general public. Governments at national, district, and local levels define and implement regulatory programs; they can also play an important role in providing incentives, removing obstacles, and creating a climate that supports effective integrated water management. Local and international financial institutions play a key role in providing resources to bring about the necessary actions and improvements, and even in catalyzing change through the stimulation of new ways of thinking about problems in the basin and their solution.

2.1. Identification of the Relevant Institutions and Actors

2.1.1. Sectors

Practices and policies in different sectors can be a cause of environmental problems or a constraint to effective action. Some of the sources of the pollution problems and water quantity problems can result from the activities of the following entities and activities.

a) Cities

Many water management problems occur close to, or downstream of, urban population and industrial centers. Municipal waste water, storm water, and seepage from improperly stored materials are important sources of surface and groundwater pollution. Wastewater is a particularly important source of organic materials, nutrients, and microbiological water pollution.

b) Rural Towns and Villages

In rural areas, where infrastructure is rarely as well developed as in urban areas, water supply and sewerage and wastewater treatment facilities are usually small and less well maintained. Abstraction of water for drinking is often from groundwater sources such as shallow hand-dug wells that are at risk from microbiological contamination derived from cesspits and leaking sewers. The lack of effective sewerage in small towns and villages also threatens shallow ground and surface sources of drinking water. Any uncontrolled solid waste disposal or poorly maintained storage of toxic and hazardous chemicals places groundwater at further risk of contamination.

c) Industry, energy production, and transport

The chemical, fertilizer, mining, food processing, metallurgy, tanning, and pulp and paper industries create serious problems. River transport contributes to oil pollution. Coal-fired power plants are sources of pollutants from flue gases and ash. Many of the hydraulic structures built for navigation and hydropower generation cause environmental problems such as those identified earlier.

d) *Agriculture*

This sector is an important source of pollution of surface and groundwater, through fertilizer, agrochemicals, and salt leached from soils. Pollution from manure from large livestock farms is a particularly serious problem. Improper cultivation practices cause erosion and areal runoff. Irrigation places heavy demands on water supply and drainage causes alterations in flows and water tables.

2.1.2. Actors

In all sectors, the key actors for change are the public authorities, public and private enterprises, NGOs, and the general public, both as citizens and as consumers. There are key relationships between these actors and the principal sectors and sources of pollution of the Danube River Basin.

a) *Public Authorities*

Governments at national, district, and local levels define and implement regulatory programs; they can play an important role in providing incentives, removing obstacles, and creating a climate that supports effective integrated water management.

The influence in the region, especially in the Lower Basin, of the European Union and of its future expansion cannot be underestimated. The key role played by the European Commission in the dispute between Hungary and Czechoslovakia/Slovakia has already been noted. As in that case, it is likely that Lower Basin countries will exercise their best efforts to bring their environmental laws and practices—including those relating to the Danube—into line with those of the EU in order to position themselves for eventual membership in the Union.

Finally, the role of international financial institutions in the future development and management of the Danube Basin cannot be ignored. Countries in the Lower Basin will need considerable financial assistance if they are to protect the ecosystems of the Danube while developing economically. Support will be needed not only for water treatment facilities and for other works, but also for capacity building in the field of environmental protection. The bulk of the support for the former is likely to come from private investment but the latter countries often rely upon the international and bilateral donor community.

b) *Public and Private Enterprises*

Under the former economic system in the Central and Eastern European countries, most enterprises were government-owned. While many of these entities are being privatized, the process is proceeding slowly. Most power generation enterprises, water utilities, and heavy industries are still owned by governments. This will doubtless change to some extent over time, due to the influence of market forces, loan conditions, and efforts of Lower Basin countries to position themselves for eventual EU membership.

c) *The General Public and Nongovernmental Organizations*

Every person in the Danube River Basin has several possible roles—as a consumer of goods and services; as a producer of waste at home at the workplace; as a user of recreation facilities; and as a citizen whose choices and actions express cultural, social, aesthetic, spiritual, and environmental values. NGOs are established by members of the public to promote public and governmental awareness of environmental issues and to assist in the development of appropriate policies.

The Gabčíkovo-Nagymaros case again provides an illustration of the potential significance of action by civil society. It will be recalled that there, Hungarian citizen protests against the project being constructed by Hungary and Czechoslovakia on the Danube were an important factor contributing to Hungary's eventual decision to withdraw from the project. The protests were focused on the project's effects on the environment. It seems likely that institutions of civil society will continue to develop in the Lower Basin countries and will have an increasing impact upon governmental decision making in relation to the sustainable development of the Danube Basin.

2.2 Relationships and Interconnections between the Different Institutions

Various Danube countries are parties to international agreements in addition to the 1994 Danube River Protection Convention that directly or indirectly affect the management of the river basin. These agreements include the following.

Europe Association agreements have been concluded between the EU and Bulgaria, the Czech Republic, Hungary, Romania, and Slovakia to prepare for their eventual accession to the European Union. A similar agreement is envisaged with Slovenia. The agreements provide for increased cooperation in a number of areas, including the environment.

In April 1993, the Environmental Action Program for Central and Eastern Europe (EAP) was endorsed by the Environment Ministers of the region of the UN Economic Commission for Europe (ECE) and by the Environment Commissioner of the EU. The EAP represents a broad consensus on environment and development. It calls for government action in three areas: first, the integration of environmental considerations into the process of economic reconstruction to ensure sustainable development; second, government action concerning institutional capacity building, including improvement of legal and administrative systems as well as management, training, and education; and third, government assistance that brings immediate or short-term relief to regions where human health or natural ecosystems are severely jeopardized, and taking into account transboundary environmental problems. Furthermore, the EAP offers illustrative investment projects for priority needs.

Launched along with the EAP in 1991 was the Environmental Programme for the Danube River Basin. This program is concerned with environmental protection and sustainable development within the Danube Basin. Like the EAP, the programme is supported by the international donor community and is Western in orientation, in the sense of incorporating modern concepts of participatory governance.

The Convention on the Protection and Use of Transboundary Watercourses and International Lakes was adopted by the Senior Advisers to the ECE Governments on Environmental and Water Problems in March 1992, and was signed by twenty-two countries and by the European Community (EC). This convention seeks to strengthen the protection and ecologically sound management of surface waters and groundwaters by providing a framework for regional cooperation on transboundary problems.

The Convention on the Protection of the Black Sea against pollution (Black Sea Convention), establishes a common legal regime for controlling marine pollution in the Black Sea. It was signed by the six Black Sea coastal countries in April 1992, and came into force in February 1994. The convention contains a legal framework for the establishment of a Black Sea Commission, and provides protocols for protection against land-based sources of pollution, for regulating dumping, and for emergency response in the case of spills.

The Convention on Wetlands of International Importance, Especially as Wild Waterfowl Habitat (the Ramsar Convention) was signed in Ramsar, Iran, in 1971. The agreement sets out measures for the protection of wetlands, particularly those that are important waterfowl habitats. It introduces the principle of the "wise use" of all wetlands, not only those listed under the convention: "wise use of wetlands is their sustainable utilization for the benefit of mankind in a way compatible with the maintenance of the natural properties of the ecosystem." One of the key elements of the convention is development and implementation of national wetland policies as tools for the delivery of wise use. This convention has been ratified by all of the Danube States except Moldova and Ukraine.

The UN Convention on Biological Diversity was signed by 72 countries and by the EU at the UN Conference on Environment and Development in Rio de Janeiro in July 1992. (The convention remained open for signature until June 4, 1993, by which time 168 States had signed it.) The convention entered into force on December 29, 1993. There are currently 177 parties to the convention, including some 8 Danube States. Its objectives are the conservation of biological diversity, the sustainable use of natural resources, and the fair and equitable sharing of the benefits of the use of genetic resources. The treaty enshrines the principle that States have the sovereign right to exploit their own resources according to their own environmental policies and the responsibility to ensure that activities within their jurisdiction or control do not damage the environment of other States or areas beyond the limits of national jurisdiction.

The tendency toward decentralization in the basin, reflected in the breakup of the Soviet Union, the "Velvet Divorce" between Slovakia and the Czech Republic, and the disintegration of the former Yugoslavia, is countered somewhat by the desire of the former socialist countries to become members of the European Union. Despite commentators' doubts about whether all countries in the basin will ultimately join the EU, this powerful unifying force will doubtless assist in forging relationships between the different institutions and legal regimes just mentioned, many of whose fields of competence already overlap.

2.3. Qualification of These Links

Links between the existing institutions and regimes are largely nonexistent, at least in the formal sense. The establishment of the Danube Protection Convention's Permanent Secretariat should assist in the building of relationships, however, since it has the capacity to function both as a clearinghouse for Danube Basin States and as a focal point for communications from external institutions. In addition, the Danube Strategic Action Program (SAP) affirmatively supports the implementation of the Danube Protection Convention.

The widest gulf appears to exist between the largely Soviet-era Danube (navigation) Commission and the more modern environmental initiatives of Western countries and institutions. It is uncertain what role the Russian Federation will ultimately play in the Danube region. Its past influence in the Danube Basin—which was manifest particularly through the Danube Commission—has diminished markedly for political and economic reasons, but it remains a major regional power whose influence cannot be ignored.

3. MANAGEMENT OF RIVER BASIN

Integrated water management concerns both minimizing the conflicts between different water uses and users, and optimizing the economic, health, and environmental benefits from water resources on a sustainable basis. Some types of water use do not necessarily conflict. Other uses are mutually exclusive or lead to conflict about the quality or quantity of water available to other users. Although the surface waters of the Danube and its tributaries are constantly being renewed, user conflicts cannot be solved while the polluting emissions continue. Groundwater requires particularly careful management since the degradation or reduction of groundwater is often irreversible or requires an extremely long period, in some cases centuries, for renewal.

Use of surface waters requires maintenance of a minimum flow at all times or at least during critical periods, and maintaining such minimum flows is especially important for the protection of the integrity of aquatic ecosystems. Human activities can harm ecosystems by not maintaining such flows, as well as by discharging pollutants and waste products into the rivers or underground aquifers of the Danube River Basin. Hence, the water quality management objectives for a given river reach depend on the uses of water there. The ultimate criteria will be set according to the users with the need for the highest quality of water so as to protect the ecosystems and biodiversity.

The water issues and problems of the Danube Basin may have different regional or transboundary impacts. Some problems, like water shortage in small tributaries, heterotopic growth, oxygen depletion, and microbiological pollution, are normally confined to limited river reaches or water bodies. They may cause health, economic, or ecological conflicts of local, national, and even transboundary importance. On the other hand, the pollution load of nutrients will normally not cause severe local water quality problems. It is, however, one of the most important problems of the Danube's wetlands and the Black Sea.

Assessment of the Actual Management of the River Basin

Management of the basin, to the extent that one could speak of basin-wide management at all, had since 1948 taken the form of coordination between riparian States in the field of navigation. But as just noted, the Belgrade Convention actually vested plenary control in riparian States over sections of the river within their territories—a control qualified only by the obligation to respect freedom of navigation. However, beginning with the Bucharest Declaration in 1985, Danube Basin States resolved to engage in bilateral and multilateral action to improve water quality. This constituted a first, major step in the direction of basin-wide cooperation in the reduction and control of water pollution. But although this program led to the establishment of a monitoring system, it has been characterized as being insufficient and ineffective, due to the wide disparities in approaches and resources among basin States.[35] In fact, it has been concluded that despite the Bucharest Declaration, "at the time of the political transitions in 1989, formal protection of the Danube environment from excessive water pollution was essentially nonexistent," due at least in part to a dearth of resources in the former socialist countries.[36]

Then in 1991 the Danube States—together with their partners, the European Community (EC), UNDP, UNEP, the World Bank, WWF, and others—established the Environmental Programme for the Danube River Basin, a project that encourages public and NGO participation throughout the planning process. Cooperation was further intensified by the conclusion of the Danube Protection Convention in 1994, which as already noted, is aimed at achieving sustainable and equitable water management in the basin. It is worth recalling here that the Danube Environmental Programme's Strategic Action Plan (SAP)[37] has become an instrument supporting the implementation of the 1994 convention.

But the political upheavals of the late 1980s and early 1990s, together with the continuing economic effects of socialism in the former CMEA countries, clearly present challenges that any attempt at cooperative management will have to address. While Upper Basin countries, which have high standards of living and enjoy economic prosperity, have open, democratic societies and are chiefly concerned with environmental protection and pollution control, Lower Basin countries are in need of economic development and may be less receptive to expensive pollution control programs—such as construction of municipal and industrial water treatment plants. On the other hand, the Upper Basin countries in many ways have less incentive to control their own pollution discharges since they are carried downstream, while Middle and Lower Basin countries, which receive these wastes, may have lower pollution standards.

These factors clearly militate in favor of management of the basin as a unit. Pollution standards must be harmonized, avoiding a "race to the bottom." Danube States have, in principle, agreed to integrated management of the basin in the Danube Protection Convention. Several factors auger well for the success of this effort. The first is the very agreement itself, something that has eluded States sharing other major international watercourses. The 1994 convention, no doubt made possible in part by the desire of the former Soviet bloc states to join Western Europe—with the promise of economic development that would entail—not only memorializes values and principles on which there is general agreement, but also establishes an institution with legal capacity to assist the Danube Basin countries in realizing the objectives of the agreement. The second auspicious factor is the interest of the international donor community in the success of the broader effort to protect, preserve, and restore the environment and ecosystems of the Danube Basin, interest made evident in donor support of the Danube Environment Programme, including activities that support efforts under the 1994 convention. The third factor is that, despite their differences, the States of the Danube Basin are ultimately European States, with many of the same fundamental values and traditions. The fourth factor, related to the third, is the positive influence of the European Union, a party to the Danube Protection Convention, both through proactive efforts to improve environmental quality in the basin, and through the incentives to protect the environment created in the Middle and Lower Basin countries by the aspirations of these States to join the EU. These incentives encourage the establishment of a more open and transparent system of governance in these countries, the ratcheting up of pollution standards, and the improvement of measures for the protection of ecosystems.

Despite the many differences between the countries of the Upper and Lower basins, all Danubian countries share certain values and have agreed, in the Danube Protection Convention, on a set of fundamental principles to govern the use and protection of the Danube. These principles include the precautionary principle; the use of Best Available Techniques, or Technologies (BAT) and Best Environmental Practices (BEP) for the control of pollution; the control of pollution at its source; a principle paid by the polluter; and a commitment to regional cooperation and exchange of information among the basin States. Some of the shared goals of the basin States are reducing the negative impacts of activities in the Danube River Basin and on riverine ecosystems and on the Black Sea; maintaining and improving the availability and quality of water in the Danube River Basin; establishing control of hazards from accidental spills; and developing regional water management cooperation.

Local needs and problems will normally be most important criteria for actions and investments in each country, but by participating in the Danube Environmental Programme and by concluding the Danube

River Protection Convention, the Danube countries have also committed themselves to addressing regional and basin wide problems.

Regional cooperation means full participation in, and utilization of, regional mechanisms and structures for international cooperation, consultation, and coordination on policy and action. Water quality and quantity and the health of the river's aquatic habitat and biodiversity depend on cooperation between all of the users of water in the basin, as well as users of land whose activities impact water resources of the basin. Regional cooperation can strengthen the efforts of the Danube countries to adopt and implement legal, administrative, and technical measures to prevent and reduce transboundary impacts; to monitor water quality and resources; and to harmonize water quality standards and pollution controls.

The sharing and exchange of data and information is fundamental to regional cooperation, effective pollution control, and the understanding and solution of regional problems. The Danube countries have agreed in Article 12 of the 1994 Convention to exchange a wide range of reasonably available data and information, including both scientific data relating to the condition of water resources and legal and experiential information concerning, e.g., experience gained with the best available techniques, measures taken with regard to transboundary impacts and information concerning domestic regulations. It is, however, not clear to what extent these provisions have actually been implemented.

Cooperative efforts in the basin, at least during times of crisis, were well illustrated during the floods of August 2002 in the Danube Basin. During that emergency situation the Slovak agriculture minister, Pavel Koncos, expressly thanked the water managers in Hungary, Austria, and the Czech Republic for their cooperation in efforts to manage the situation along the Slovak section of the Danube.[38] Danube waters rose to precarious levels throughout August: floodwaters were initially predicted to crest in Slovakia at up to 500-year levels; ultimately they settled closer to 100-year levels.[39] (In the Slovak capital of Bratislava, statements of public reassurance focused upon the capability of the Gabčíkovo Dam to cope with even a thousand-year high water crest).[40]

While the Danube and the Black Sea region is an area of increasing significance in the context of an enlarged Europe, it suffers from several environmental and health problems. Additionally, the Danube-Black Sea system, which has had been called the single most important freshwater system in Europe,[41] is of a strategic importance that will only increase as the EU enlarges.[42] In the fall of 2001 the European Commission established an informal Task Force for cooperation on water-related issues in the Danube-Black Sea Region (DABLAS Task Force).

DABLAS consists of the European Commission, interested EU member States, international financing institutions, and bilateral donors.[43]

DABLAS at an informal level and the International Commission for the Protection of the Danube River established by the Danube Protection Convention[44] at a formal level provide evidence of a new twenty-first century dialogue among the Danube Basin States. This modern dialogue seeks to emphasize standardization and transparency and inclusiveness. While Danube Basin States have political borders there is increased emphasis upon the shared interest and investment in the future of the common resources of the Danube.

3.2. How the Situation is Likely to Evolve

Current trends all seem to be in the direction of enhanced cooperation and the strengthening of institutions in the Danube Basin, both of which auger well for increased protection of the Danube River Basin and its ecosystems.

There appear to be constructive synergies between the less formalized and more decentralized Danube Environmental Programme and the more traditional regime of the Danube Protection Convention. The agenda of the former has been characterized as being broader than that of the latter, since the former aims at sustainable management of the basin while the latter is concerned chiefly with prevention and reduction of pollution. "Whereas the Danube Program and accompanying Strategic Action Plan follow the governance trend in international environmental affairs characterized by 'soft law' initiatives, the absence of formal legal instruments and broadly-based networks, the Danube Convention follows the tried-and-tested environmental law approach to water management cooperation by establishing a formal decision hierarchy."[45] The Danube River Basin Environmental Declaration, adopted in Bucharest in December 1994, endorsed the approach of the Danube Strategic Action Plan.

These synergies between the two initiatives should continue to strengthen efforts to protect the environment of the Danube Basin. However, it appears that the road to adequate environmental protection and sustainable development within the basin will not be a short one, in view of the tremendous changes that have taken place within the basin in the past decade. The former socialist countries are not only relatively poor, in comparison with the Upper Danube States, but also lack the kind of historical culture of participatory democracy that is prevalent in the West. But between the "grassroots" approach of the SAP, the more centralized system of the 1994 convention, and the positive influence of the European Union in the basin, the necessary processes, incentives, and institutions appear to be in place to permit and guide steady progress along that challenging road.

3.2. Conclusion: What Should Be Done about It?

It would appear that in view of the existence of a number of institutions and programs that are solely or partly concerned with the Danube Basin, the current need is for coordination. This is particularly true of the venerable but somewhat archaic Danube Commission, whose functions should be integrated into the regime of environmental governance of the basin. While not all aspects of navigation are related to nonnavigational uses of the Danube and its tributaries, navigational uses can affect the aquatic environment and nonnavigational uses can impact navigation. For these reasons alone, coordination of the work of the various institutions and programs would be desirable, if not necessary.

Perhaps the greatest need in the basin appears to be funding for the projects necessary to effect improvement in the water quality and environment of the Danube Basin. Mobilizing and effectively delivering funding will not be an easy task, since many of the kinds of projects that must be undertaken are not of the "showcase" variety that often attract major donors. Furthermore, some donors may be hesitant to channel funding for environmental or pollution-control projects through governments of former socialist countries, many of which would be challenged to effectively administer such grants or loans. But the existence of the International Commission for the Protection of the Danube River should help to alleviate many of these problems, because it can act as a clearinghouse for projects and for the receipt and administration of project funding. This, however, points to the necessity of ensuring that the commission itself is adequately staffed and funded, and remains so over time. The participation of the EU as a party to the Danube Protection Convention, together with its broader influence in the region, should help to both ensure a strong and viable Danube Protection Commission and enhance coordination between the various programs and institutions in the region.

NOTES

1. Charles the Great had first ordered the construction of such a canal in 793 AD but the project was not completed. King Ludwig I of Bavaria launched a successful project, which linked the two basins in 1846 but was only suitable for small ships.

2. Some estimates are larger. For example, Kliot, Shmueli, & Shamir give a figure of ca. 828,000 km². N. Kliot, D. Shmueli & Uri Shamir, *Institutional Frame-*

works for Management of Transboundary Water Resources, p. 192 (2 vols., Water Research Institute 1997).

3. See, e.g., "Thousands Flee Prague Flooding; Heavy Rains Persist across Central Europe, Causing Flooding and Dozens of Deaths," *St. Petersburg Times*, 14 Aug. 2002, available at <http://www.sptimes.com/2002/08/14/Worldandnation/Thousands_flee_Prague.shtml>.

4. Kliot, Shmueli, & Shamir (1997). *supra* note 2, at p. 196.

5. World Conservation Union (IUCN), *Analysis and Synthesis of National Reviews for the Danube River Basin Environmental Programme*. Final Report, p. 20 (1994).

6. See generally Stephen Gorove, *Law and Politics of the Danube: An Interdisciplinary Study* (The Hague: Martinus Nijhoff, 1964). See also G. Kaeckenbeeck, *International Rivers*, secs. 98–180 (1918, reprinted, New York: Oceana and London: Wildy & Sons 1962); and L. A. Teclaff, "Fiat or Custom: The Checkered Development of International Water Law," 31 *Natural Resources Journal* 44 (1991).

7. P. Reuter, *International Institutions*, p. 207 (New York 1961).

8. 30 March 1856 (Austria, Britain, France, Ottoman Empire, Prussia, Russia, and Sardinia), 114 Parry's T.S. 409. The Treaty of Paris ended the Crimean War.

9. Treaty of Peace between the Allied and Associated Powers and Hungary, June 4, 1920, S. Exec. Doc. 3539.

10. Ibid., art. 293.

11. Ibid., art. 282.

12. 33 UNTS 181 (USSR, Bulgaria, Czechoslovakia, Hungary, Romania, Ukraine, & Yugoslavia).

13. Teclaff, *supra* note 6, at p. 56.

14. In 1977 the World Health Organization warned that the Danube suffered from inadequate pollution control.

15. Reprinted in Aktuelle Österreichische Praxis zum Völkerrecht 1985–1986, *Österreichische Zeitschrift für öffentliches Recht und Völkerrecht* 429 (P. Fischer & G. Hafner, eds., 1986).

16. EU Official Journal No. L 342, 12/12/1997 P. 0019–0043. Also available at <http://ksh.fgg.uni-lj.si/danube//envconv/>.

17. Case concerning the Gabcikovo-Nagymaros Project, 1997 ICJ Rep. 7.

18. Treaty concerning the Construction and Operation of the Gabcikovo-Nagymaros System of Locks, 16 Sept. 1977, 32 ILM 1247 (1993) (hereafter "1977 treaty").

19. 1997 ICJ Rep. paragraph 15, at p. 18.

20. 1977 treaty, art. 1(4).

21. 1977 treaty, art. 1.

22. Variant C was one of the alternative solutions considered by Czechoslovakia.

23. Special Agreement for Submission to the International Court of Justice of the Differences between the Republic of Hungary and the Slovak Republic concerning the Gabcikovo-Nagymaros Project, Brussels, 7 April 1993, 32 ILM 1291 (1993).

24. The ICJ ruled that the political conditions prevailing at the time the treaty was concluded neither constituted an essential basis of the parties' consent nor, when they changed, radically altered the extent of the obligations still to be performed. 1997 ICJ Rep. paragraph 104, at p. at 65.

25. 1997 ICJ Rep. paragraph 17, at pp. 18–20.

26. Kliot, Shmuel, & Shamir, *supra* note 2, at p. 196.

27. See generally Aaron Schwabach, "The Tisza Cyanide Disaster and International Law," 30 *Environmental Law Report* 10509 (2000).

28. J. Linnerooth-Bayer & S. Murcott, "The Danube River Basin: International Cooperation or Sustainable Development," 36 *Natural Resource Journal* 521, 532 (1996).

29. Ibid., pp. 531–532. Additional detail concerning forms and levels of surface and ground water pollution is contained in ibid., p. 532. See also *Environmental Programme for the Danube River Basin, Strategic Action Plan for the Danube River Basin* 1995–2005 (1994).

30. Linnerooth-Bayer & Murcott, *supra* note 28, at p. 529; Kliot, Shmueli, & Shamir *supra* note 2, at p. 197.

31. See 1997 *I.C.J. Reports* p. 7.

32. Kliot, Shmueli, & Shamir *supra* note 2, at p. 199.

33. Ibid. See also Linnerooth-Bayer, "The Danube River Basin: Negotiating Settlements to Transboundary Environmental Issues," 30 *Natural Resource Journal* 629, at p. 632 (1990).

34. Linnerooth-Bayer & Murcott, *supra* note 18, at p. 522.

35. IUCN, *Final Report, supra* note 4, at p. 75.

36. Linnerooth-Bayer & Murcott, *supra* note 18, at p. 537, citing Westing, A. H. "Environmental Security for the Danube River Basin," 16 *Environmental Conservation* 327 (1989).

37. Note 19, *supra.*

38. *Situation in Bratislava Is Getting Worse but Not Dramatically.* Czech News Agency, CTK National News Wire, 15 August 2002.

39. Ibid.

40. Ibid.

41. Ibid.

42. Commission of the European Communities, Press Release, IP: 01/1651, *Unprecedented Action to Protect the Danube River and the Black Sea,* 23 November, 2001.

43. Commission of the European Communities, Press Release, *Commission Gives New Impetus to Environmental Co-operation in the Danube-Black Sea Region,* 31 October, 2001.

44. Danube Protection Convention Article 18 and Annex IV.

45. Linnerooth-Bayer & Murcott, *supra* note 18, at p. 544.

Bibliography

Commission of the European Communities, Press Release (2001), IP: 01/1651. *Unprecedented Action to Protect the Danube River and the Black Sea,* November 23.

Commission of the European Communities. Press Release (2001), *Commission Gives New Impetus to Environmental Co-operation in the Danube-Black Sea Region*, October 31.

Gorove, S. (1964), *Law and Politics of the Danube: An Interdisciplinary Study* (The Hague: Martinus Nijhoff).

Kaeckenbeeck, G. (1962), *International Rivers*, secs. 98–180 (1918, reprinted, New York: Oceana and London: Wildy & Sons).

Kliot, N., D. Shmueli & U. Shamir (1997), *Institutional Frameworks for Management of Transboundary Water Resources* 192 (2 vols., Water Research Institute).

Linnerooth-Bayer, J. & S. Murcott (1996), "The Danube River Basin: International Cooperation or Sustainable Development," 36 *Natural Resources Journal*.

Linnerooth, J. (1990), "The Danube River Basin: Negotiating Settlements to Transboundary Environmental Issues," 30 *Natural Resources Journal*.

"Thousands Flee Prague Flooding; Heavy Rains Persist across Central Europe, Causing Flooding and Dozens of Deaths, *St. Petersburg Times*, August 14, 2002, available at http://www.sptimes.com/2002/08/14/Worldandnation/Thousands_flee_Prague.shtml.

Reuter, P. (1961), *International Institutions* (New York).

Schwabach, A. (2000), "The Tisza Cyanide Disaster and International Law," 30 *Environmental Law Report*.

Special Agreement for Submission to the International Court of Justice of the Differences between the Republic of Hungary and the Slovak Republic concerning the Gabcikovo-Nagymaros Project, Brussels, 7 April 1993, 32 ILM 1291 (1993).

Teclaff, L. A. (1991), "Fiat or Custom: The Checkered Development of International Water Law," 31 *Natural Resources Journal*.

Westing, A.H. (1989), "Environmental Security for the Danube River Basin," 16 *Environmental Conservation*.

World Conservation Union (IUCN) (1994), *Analysis and Synthesis of National Reviews for the Danube River Basin Environmental Programme*, Final Report.

Chapter 4

The Euphrates River Watershed: Integration, Coordination, or Separation?

Arnon Medzini and Aaron T. Wolf

1. THE GENERAL CONTEXT

1.1. The Biophysical Dimension

The Euphrates River rises north of Erzurum in Turkey, where the highest mountains are more than 3,000 m above sea level. (See Figure 6 and Table 9.) The sources are situated between the Black Sea and Lake Van and the river is formed by the confluence of two tributaries, the Murad-Sue, which originates in the many springs in the area of Ala-Dag and flows from the south, and the Kara-Sue, which begins in the Kargapazari Mountains, north of Erzurum and flows from the north to meet at a point near the village of Keban. The Murad-Sue River is 650 km long and the Kara-Sue River is 450 km long. These tributaries contribute 75 percent of the Euphrates water. From 1,000 m, above sea level in north Anatolia, the river is then known as the Euphrates.[1]

Ten kilometers downstream from the confluence, there is a narrow gorge where the Keban Dam has been constructed. Downstream from the dam many tributaries and wadis join the river, increasing its discharge. The most important tributary is the Tokhma Sue that drains the Taurus Mountains and joins the Euphrates near the town of Malorteya. Below the village of Keban, the Euphrates flows down the steep slopes on the southern margin of the mountains of the Kurdistan and Armenian areas and enters the region of the eastern plains of Syria. The total length of the Euphrates in Turkey is 455 km, from the point of confluence of the Kara-Sue and Murad-Sue until it enters Syria at the town of Trablus.

The Euphrates has three tributaries in Syria: the Sajur joins the Euphrates on the right bank about 30 km downstream from Jerablus; the Balikh joins on the left bank, halfway between Jerablus and Dier-ez-Zor; below Dier-ez-Zor, the river broadens out somewhat and contains numerous rocky stretches, shallow rapids, and islands and is joined by the last important tributary, the Khabur, about 30 km below Deir-ez-Zor. The Khabur rises on the south-facing slopes of the Taurus Mountains in Turkey and drains into northeastern Syria. South of the Khabur River the Euphrates River receives no further discharge from any source, but a number of dry riverbeds indicate that in relatively recent geologic time a number of tributaries flowed into the Euphrates from what are now the Syrian and Arabian deserts. The total length of the Euphrates in Syria is 675 km and it enters Iraq at the Husaiba Settlement, after flowing through Abu Kamal in Syria.

Having crossed the Iraqi border, the Euphrates continues its flow in a southeasterly direction, crossing the desert uplands and narrow wadis that end at Ramadi before entering the Mesopotamian plains. As there is no great difference in elevation between the riverbeds and the surrounding irrigated areas, gravity canals are used to divert the water for irrigation.[2] Although the Euphrates comes very close to the Tigris near the town of Falluja, it again flows away from it in a southern direction. Since the water level in the Euphrates is higher there than that of the Tigris at a similar point, the land between the two rivers can be irrigated by diverting Euphrates water from its left bank by means of canals.

Downstream from the town of Mussaiyab and upstream of the Hindiya barrage many large canals draw water from the Euphrates River. The most important is Shatt-al-Hilla (called Shatt-al-Hindiya at this point), which passes the Hindiya barrage and then flows through the towns of Hindiya and Kifil. Downstream from Kifil it bifurcates into two branches: the Kufa branch on the west and the Shamiya branch on the east. The Kufa branch again bifurcates downstream from the town of Abu Shukhir into the Jehad and Mishkhab branches while the Jehad branch ends in an area of agricultural land where its waters are completely utilized for irrigation purposes. The Mishkhab branch (of the Kufa branch) continues to flow in a southern direction and passes through Mishkhab, Qadisiya, and then to the south of Shinafiya. Here the river then flows into many branches that join one another to the north of the town of Nasiriya. From Nasiriya the river again flows into many branches toward the town of Souk-al-Shiukh and all of these branches discharge into the Hammar Marshes. From the Hammar Marshes the Euphrates waters flow into Shatt-al-Arab, near Karmat-Ali.

The total length of the Euphrates in Iraq is about 1200 km and the total length of the Euphrates from the confluence of the Kara-Sue and Murad-Sue to its confluence with Shatt-al-Arab near Karmat-Ali is about 2,330 km. The Euphrates and its tributaries drain a basin of 444,000 km² in area, 28% of which lies in Turkey, 17% in Syria, 40% in Iraq, and 15% in Saudi Arabia.[3] In its course, the river flows through four topographic regions:[4]

a) The high mountainous region. This is located in the northern part of the basin, from the sources of the river to the area south of the village Cungus in Turkey. Here the region rises to an altitude of 1,500 to 3,000 m above sea level, where the river flows through narrow and deep wadis and later through a line of basins and wadis until it reaches the Anti-Taurus system. In this region, the river flows rapidly and is characterized as an alpine river.

b) The foothill region and the Kracali heights. This region is located south of the mountainous region south to Chingiz (740 m) and Birecik (420 m) in southern Turkey. This is the transition zone between the mountains and the lowlands and most of this area is located in Turkey. The Euphrates flows in a cleft that is partly covered with lava accumulations and the river cuts down into a deep and curved valley 20–30 m beneath the surrounding area. Here the river's stream becomes wider and its flow weakens.

c) The Plains of Gezira. These are located south of the foothills stretch, from Birecik (420 m) to Hit in Iraq (52 m). In this area the river flows in a series of extensive flood plains. The river meanders through a wide valley and geologic circumstances cause the river to flow between elevated bluffs

d) The Plains of Iraq and the delta from Hit until the Euphrates joins the Tigris River. This region is a flat, delta-like plain. The river flows across an alluvial plain built up from eroded material upstream. The river gradient is small (sometimes no more than 510 cm per km). In the plains region, there are wide areas of marshy lands including streams and meanders that have been abandoned.

The Euphrates River, from its sources in northeast Turkey up to its confluence at Shatt al-Arab, flows through three climate zones that differ significantly from one another:

a) The Mountainous Mediterranean Climate Zone. The climate here determines the flow regime of the river. Temperatures in the mountains frequently fall bellow 0°C during the winter and

the Euphrates is largely fed by precipitation falling over the uplands of eastern Turkey, where the annual total precipitation often
exceeds 1,000 m. As most of this precipitation occurs during the
winter months as snow, it tends to be locked up as snow and ice,
but with rising temperatures in spring and early summer, the snow
fields melt. The rainy season starts in October and ends in April.[5]

b) The Semi-Arid Mediterranean-type Zone. This zone includes those
   areas where there is a small winter water surplus. Such areas can
   be described as having a steppe type vegetation, climate, and land
   use. Temperatures in the winter months in Urfa, Turkey, a site
   with a climate typical of the foothills region, range from 5°C to
   7°C and in the summer, they range from 27°C to 30°C. The average rainfall is 300 m per annum near the Turkish-Syrian border.[6]

c) The Arid Zone of South Syria and Iraq. Rainfall in the plains is
   characterized by a low average precipitation of 150–200 m per
   annum that occurs mainly in the November-April winter season.
   The rainfall is not reliable in any part of the plains and the records
   show large fluctuations from year to year. Summer in this region
   is intensely hot, with day shade temperatures frequently reaching
   a maximum of 45°C in July and August and from 30°C to 35°C
   in the Al Jazira subregion. Throughout the entire basin the winter season (December, January, and February) is the most humid,
   with over half of the annual precipitation in the valley falling
   during these months. In comparison, the summer season is very
   dry and brings little precipitation.

Because of this topographic and climatic variability, the amount of
water in the Euphrates River, the third largest river in the Middle East
after the Nile and the Tigris, varies considerably from month to month
and from year to year (Figures 6 and 7). The annual cycle of the
Euphrates River discharge can be divided into three parts:

a. Period of high discharge—March to June
b. Period of low discharge—July to October
c. Period of average discharge—November to February

The melting of the winter snow in the uplands of Turkey releases large
quantities of water into the river to produce a discharge peak during
April and May, when the discharge at Hit averages 2,400 m³/sec. The
discharge drops sharply in June and July as the frozen precipitation of
the winter season is exhausted and as the nearly rainless summer begins.

In August, September, and October the mean discharge at Hit is
around 300 m³/sec. In a year of heavy winter precipitation, however, the

peak discharge in May can reach 4,300 m³/sec, while there is little increase in the average summer flow level. After a dry winter, the discharge in April or May can be as low as 1,300 m³/sec and may drop to about 100 m³/sec in August and September. Thus, in one year, as much as twice the average amount of water may flow in the Euphrates while, in another, little more than half of the average annual discharge will be generated. The discharge of the Euphrates diminishes systematically with distance downstream after its confluence with the Khabur, primarily as a result of evaporation and infiltration into the subsurface. In the swampland of the upper delta region, both before and after confluence with the Tigris, the influence of large-scale transpiration by aquatic vegetation further diminishes the flow.[7]

The seasonal distribution of the water supply of the system does not coincide with crop needs. Winter crops in the riparian countries need water from May onward, but the low-water season lasts from July to December and, during this period, the mean water discharge of the Euphrates is 421 m³/sec. Thus the river reaches its lowest levels in September and October when water is badly needed. When the crops are either half-grown or almost ready for harvest and water is no longer needed, the fields are subject to the danger of inundation since spring is the flood season. During this period, the mean water discharge of the river is 1,765 m³/sec. In the case of summer crops, which need water from April to September, the situation is reversed. At first they receive abundant supplies, and then the supply declines gradually until it reaches a low point in September.[8] The normal difference between high and low levels is about 3.3 m.[9]

1.2. The Historical Context

Because of water's preeminent role in survival—from an individual's biology to a nation's economy—political conflicts over international water resources tend to be particularly contentious. The intensity of a water conflict can be exacerbated by a number of factors, including a region's geographic, geopolitic, or hydropolitic landscape. Water conflicts are especially bitter, for example, where the climate is arid, where the co-riparians of regional waterways are otherwise engaged in political confrontation, or where the population's water demand is already approaching or surpassing its annual supply—in other words, in basins precisely like the Euphrates.

Living as they do in a transition zone between Mediterranean subtropical and arid climates, the people in and around the major watersheds of the Middle East have always been aware of the limits imposed by scarce water resources. Settlements sprang up in fertile

valleys or near large, permanent wells, and trade routes were established from oasis to oasis. In ancient times, cycles of weather patterns
had occasionally profound effects on the course of history. For example,
recent research suggests that climatic changes 10,000 years ago, which
caused the average weather patterns around the Dead Sea to become
warmer and drier, may have been an important factor in the birth of
agriculture in the region.[10]

The fluctuating waters of the ancient Euphrates have given rise to
legend, extensive water law, and to the roots of modern hydrology: The
flood experienced by Noah is thought to have centered its devastation
around the Babylonian city of Ur, submerging the southern part of the
Euphrates for about 150 days, while the code of King Hammurabi
contains as many as 300 sections dealing with irrigation. The practice
of field surveying was invented to help harness the flooding Nile.[11] And
the first, and to date the only, actual "water war" occurred along the
Tigris Basin—between the city-states of Lagash and Umma, over 4,500
years ago![12]

In the centuries since, the inhabitants of the region and the con-
quering nations that came and went have lived mostly within the limits
of their water resources, using combinations of surface water and well
water for survival and livelihood.[13] It is in the beginning of this century,
as the competing nationalisms of the region's inhabitants began to re-
emerge on the ruins of first the Ottoman, then the British, empires, that
the quest for resources took on a new and vital dimension, and by the
last quarter of the twentieth century, unilateral water developments
came very close to leading to warfare along the Euphrates.

The three riparians to the river—Turkey, Syria, and Iraq—had
been coexisting with varying degrees of hydropolitical tension through
the 1960s. At that time, population pressures drove unilateral developments, particularly in southern Anatolia, with the Keban Dam (1965–
1973), and in Syria, with the Tabqa Dam (1968–1973).[14]

Bilateral and tripartite meetings, occasionally with Soviet involvement, had been carried out between the three riparians since the mid-
1960s, although no formal agreements had been reached by the time the
Keban and Tabqa dams began to fill late in 1973, resulting in decreased
flow downstream. In mid-1974, Syria agreed to an Iraqi request that
Syria allow an additional flow of 200 Mm³/yr. from Tabqa. The following year, however, the Iraqis claimed that the flow had dropped from
the normal 920 m³/sec to an "intolerable" 197 m³/sec, and asked that
the Arab League intervene. The Syrians claimed that less than half of
the river's normal flow had reached its borders that year and, that after
a barrage of mutually hostile statements, pulled out of an Arab League

technical committee formed to mediate the conflict. In May 1975, Syria closed its airspace to Iraqi flights and both Syria and Iraq reportedly transferred troops to their mutual border. Only mediation on the part of Saudi Arabia was able to break the increasing tension, and on June 3, the parties arrived at an agreement that averted the impending violence. Although the terms of the agreement were not made public, T. Naff and R. C. Matson[15] cite Iraqi sources as privately stating that the agreement called for Syria to keep 40% of the flow of the Euphrates within it borders, and to allow the remaining 60% through to Iraq.

The Turkish GAP project has given a sense of urgency to resolving allocation issues on the Euphrates. The Southeast Anatolia Development Project (GAP is the Turkish acronym) is a massive undertaking for energy and agricultural development which, when completed, will include the construction of 21 dams and 19 hydroelectric plants on both the Tigris and the Euphrates. Land in the amount of 1.65 million hectares are to be irrigated and 26 billion kWh will be generated annually with an installed capacity of 7,500 MW. If completed as planned, GAP could significantly reduce downstream water quantity and quality.

A Protocol of the Joint Economic Committee was established between Turkey and Iraq in 1980, which allowed for the creation of Joint Technical Committee meetings relating to water resources. Syria began participating in 1983, although meetings have been intermittent at best.

A 1987 visit to Damascus by Turkish Prime Minister Turgut Ozal reportedly resulted in a signed agreement for the Turks to guarantee a minimum flow of 500 m³/s across the border with Syria. According to J. F. Kolars and W. A. Mitchell,[16] this total of 16 billion m³/yr. is in accordance with prior Syrian requests. However, according to Naff and Matson,[17] this is also the amount that Iraq insisted on in 1967, leaving a potential shortfall. A tripartite meeting between Turkish, Syrian, and Iraqi ministers was held in November 1986, but yielded few results.[18]

Talks between the three countries were held again in January 1990, when Turkey closed the gates to the reservoir on the Ataturk Dam, the largest of the GAP dams, essentially shutting off the flow of the Euphrates for thirty days. At this meeting, Iraq again insisted that a flow of 500 m³/sec cross the Syrian-Iraqi border. The Turkish representatives responded that this was a technical issue rather than one of politics and the meetings stalled. The Gulf War that broke out later that month precluded additional negotiations.[19]

In their first meeting after the war, Turkish, Syrian, and Iraqi water officials convened in Damascus in September 1992, but broke up after Turkey rejected an Iraqi request that flow crossing the Turkish border be increased from 500 m³/sec to 700 m³/sec.[20] In bilateral talks in January 1993, however, Turkish Prime Minister Demirel and Syrian President Assad discussed a range of issues intended to improve relations between the two countries. Regarding the water conflict, the two agreed to resolve the issue of allocations by the end of 1993. Although this has not, to date, been done, Demirel declared at a press conference closing the summit that, "There is no need for Syria to be anxious about the water issue. The waters of the Euphrates will flow to that country whether there is an agreement or not."[21]

1.3. The Stakes

A study of the conflict between the States concerning the sharing of Euphrates River water shows that the subject of water cannot be isolated from broader geographic, historical, political, and economic issues. Examination of the relations among the riparian states shows the importance of four principle factors: the respective policies of the three States toward the amount of water to be drawn from the Euphrates and Tigris rivers, the Kurdish question, the rivalry between the Iraqi and Syrian branches of the Ba'ath Party, and Syria and Turkey's historical animosity.

1.3.1. Relations between Turkey and Syria

a) Hydrostrategic Territory

The source of the difficulties in achieving a real improvement in relations between Turkey and Syria up to 1991 undoubtedly lies in the dichotomy of Turkey's membership in the North Atlantic Treaty Organization (NATO) and Syria's historic reliance on support from the former Soviet Union. In other words, the confrontation between East and West is reflected in the Middle East in general and in Turkish-Syrian relations in particular. Even though this tension has been reduced by the worldwide softening of the cold war, it has not completely disappeared.

Relations between the two countries have not been cordial since 1939, when France, then the mandatory power in Syria, handed the area around Alexandretta (Iskanderun in Turkish, Hatay in Arabic) over to Turkey as a bribe to enter World War II on the side of the Allies. Turkey accepted but then stayed neutral. Syria has never accepted this territorial loss and Syrian maps still show the territory as part of Syria. Damascus has never been able to hide the fact that it considers Turkish sovereignty over the Hatay as illegitimate.

In terms of water, difficulties between Syria and Turkey on the Euphrates are tied to relations on the Orontes as well. The Orontes River emanates from Lebanon, passes through Syria, and flows into the

Mediterranean Sea from the Turkish province of Hatay. It covers 40 km in Lebanon, 120 km in Syria, and 88 km in Turkey. Lebanon has two regulators on the Orontes and Syria has two dams, namely Destan and Maherde, in addition to a water regulator in the town of Jisr-Al-Sughur.

Both Lebanon and especially Syria intensively utilize the Orontes for irrigation purposes. Syria makes use of 90 percent of the river's total flow, which reaches an annual average of 1.2 billion m^3 at the Turkey-Syria border. Therefore, out of this total capacity, only the meager amount of 120 million m^3 enters Turkey, after having been heavily used by Syria. Even this small amount is expected to decrease to approximately 25 million m^3 when the planned reservoirs of Ziezoun and Kastoun in Syria are built in addition to the existing dams on the river.

More importantly, Syria never even bothered to inform Turkey about building the dams on the Orontes River even though Turkey is one of the downstream riparians. This insufficient amount of water allocated to Turkey has dried the once-productive Amik Valley, causing the death of thousands of animals and severely affecting the agriculture and animal husbandry in the region. Consequently, by using almost all of the river's water, Syria has inflicted great damage on Turkey and on the Turkish economy.

However, Syria and Lebanon concluded two agreements in 1972 and 1994 earmarking 80 million m^3 of water from the Orontes River for Lebanon's utilization. Again, Turkey, despite being one of the riparian countries, was excluded from these agreements and was neither consulted nor informed.

In light of these facts, many comparable lessons can be drawn between the Orontes and the Euphrates rivers with respect to both the amount of water released to the downstream countries and to its utilization. Syria accuses Turkey, the upstream country, of reducing the amount of water in the Euphrates, while in the case of the Orontes River, where Syria is the upstream country, it utilizes almost all of the river's water and releases a meager amount to its downstream country, Turkey. Not only that, but the dams on the Euphrates River, besides contributing to Turkey's irrigation needs, serve mainly to regulate the major fluctuations of this river so that the downstream neighboring countries might receive a regular and stable supply of water. These dams have been, and will continue to be built with, Syria's full knowledge on this principle.

In this respect, Syrian claims to Turkish territories should also be noted. Syria has long refused to recognize the transboundary character of the Orontes River. By claiming that the Turkish province of Hatay belongs to Syria, it argues that the river flows into the Mediterranean

Sea from Syria. Because of this claim, Syrian authorities have always re-
frained from discussing the situation of the Orontes River with Turkey.[22]

b) *Geopolitics and Hydropolitics*

Until the late 1970s, diplomatic relations between Turkey and Syria
remained "correct but strained." One major strain was the fact that
Syria was willing to permit anti-Ankara Kurds and leftist opposition
groups to use its territory as a base for operations against Turkey.
Ankara has sometimes angrily admonished Damascus on this issue.[23]
The subject of Syria's helping Turkish opposition groups during the
1970s by arming them and sending them back secretly into Turkey has
aroused the most severe displeasure in the Turkish administration. Evi-
dence has also been put forward suggesting that Syria had helped Armenian terrorists and that during the 1980s had similarly provided
arms for Kurdish separatists. In particular, Syria was held responsible
for the fact that Abdullah Ocalan, the leader of the Kurdistan Workers'
Party (PKK is the Syrian acronym) founded in Turkey in 1978, had
moved his headquarters to Damascus and from 1984, had sent Kurdish
fighters, trained in the Bekka Valley of Lebanon (under Syrian control),
into southeast Anatolia. Syria has refused to admit training and arming
members of the PKK, or providing facilities for them.

In December 1986, Turkish police claimed that they had discovered Syrian-backed terrorists planing to blow up the Ataturk dam.[24] The
Turks reported that there were three training camps in northern Syria
housing militants belonging to the "Armenian Secret Army for the Lib-
eration of Armenia" (ASALA) and the PKK, and that Syrian agents,
disguised as diplomats, had delivered arms to ASALA militants in various European countries. On September 17, 1986, the PKK Central
Committee held a meeting in Damascus where it decided on increasing
its operations, especially its crossings of the Turko-Syrian border into
Turkey.[25] Syrian Prime Minister Abed al-Rauf al-Kasm stated in 1986
that Damascus was unable to prevent PKK incursions into Turkey because Syria, with a long frontier, had no army in the north since it was
needed to watch the enemy in the south. Thus, although Syria was
doing its best to avoid friction with Turkey, it could not keep the border
under strict control.[26]

In July 1987, Turkish Prime Minister Ozal went to Damascus and
two months later Kadurra, the vice-prime minister of Syria, went to
Turkey. At these meetings, a protocol for "Cooperation on Security
Problems" was signed, bringing up a number of bilateral matters dealing with the prevention of the smuggling of goods across the common
border, the cross-border trade in counterfeit money, and the return of

fugitives to the country from which they were escaping.[27] In addition, discussions were held on the prevention of terrorism (from Turkey's point of view, the PKK, and from Syria's, certain elements of the Syrian Moslem Brotherhood). Particular attention was given to the important problem of regulating the water of the Euphrates.

As a result of his visit, Ozal proposed Turkish help in the prospecting for gas and oil in Syria, presented a project to supply electricity to Syria if Damascus needed it, and suggested that increased trade and economic cooperation would benefit both countries. His major proposal was, however, the installation of the Peace Pipelines.[28] According to Briefing (1989), the result of this protocol was that the Syrian government moved the PKK camp out of Syria into the Bekka Valley in Lebanon. Turkish officials felt that such a move was not sufficient in itself because there was evidence of Syrian territory still being used for many of the PKK attacks. They did not accept the Syrian excuse that the PKK camps were in sovereign Lebanese territory and therefore beyond their reach for, in fact, the Bekka Valley was under Syrian control. Many questions were left unanswered such as the fate of the PKK terrorists in the Bekka Valley and Syria's attitude toward the extradition of the PKK leader, Abdullah Ocalan, about whose presence in Syria Turkey had documentary proof but about whom Damascus claimed complete ignorance.[29]

In October 1998, after a strained decade of futile talks, Syria and Turkey signed a security agreement, in which Damascus pledged to stop its support of the Kurdistan Workers Party. This happened after Turkey has threatened military action against Syria, charging Damascus with supporting the PKK against Ankara and harboring Ocalan. According to the deal, Syria pledged to no longer allow PKK rebels or its leader to operate on its soil and recognized the PKK as a "terrorist organization."

Ocalan was expelled from Syria in November 1998 and then traveled from country to country in search of a safe haven until he took refugee in the Greek Embassy in Nairobi, Kenya. Here, Turkish commandos captured Ocalan in February 1999. Ocalan was sentenced to death for treason for leading the Kurdistan Workers Party in its armed struggle for Kurdish self-rule.[30] For the European Union (EU), the confirmation of the death sentence has already put a strain on relations with Turkey following a recent rapprochement. Turkey is trying to join the European Union, whose member countries have abolished the death penalty. Several Turkish politicians have said that Abdullah Ocalan should be hanged before the country abolishes the death penalty to comply with the standards of the European Union. EU countries have told Turkey that executing Ocalan could hurt its chances of acceptance.[31]

Syrian and Turkish officials have signed an agreement on cooperation in the fight against terrorism, as well as against organized crime
and drug trafficking. The Memorandum of Understanding was signed
on September 28, 2000 during a three-day visit by Syrian officials in
Turkey. This agreement follows several others between the two countries, most recently, the 1998 Adana Agreement, intended to diminish
the PKK's abilities to operate from Syria and Lebanon. Syria has a
record of violated territorial obligations with Turkey, which has occasionally led to tension along the 800-km-long Turkish-Syrian border.

The agreement signed with Syria seems to be part of a larger effort
to create a cooperative environment between Turkey and its neighboring countries.

The matter of counterterrorism cooperation is now engaging the
attention of top Turkish officials. This seems to be a reflection of the
understanding that cooperation is a more efficient way of combating
terrorism, especially that operating across common borders. In addition, efforts are being made to show Turkey as a calmer and safer
place—a "European" country, rather than a "Mediterranean" country.

Following the agreement, the two states began improving relations
by holding a series of talks on several bilateral issues including water
rights. Bashar Al-Assad became Syrian president following his father's
death in June 2000. He has continued his father's decision to improve
relations with Ankara, and these improvements are now leading to a
growth in Turko-Syrian business links. Another indication of how rapidly Turkey and Syria have moved to "normalization" is the following:
A report in *Al-Hayat* (London) on May 2, 2001 cited diplomatic sources
in Damascus as saying that Turkey has offered Syria a proposal for
cooperation in military training between the two countries. On May 1,
Al-Hayat quoted the Turkish ambassador in Damascus, as saying that
Syria and Turkey have "begun to cooperate in the field of military
training." ·He also stated that the two countries are negotiating the
terms of a bilateral declaration of principles "organizing the relationship between the two sides." He added that the two sides have resolved
their border conflict over the Turkish province of Hatay.[32] In August
2001, Syria and Turkey signed an agreement to cooperate on GAP,
although Turkey made no specific commitments regarding the amount
of water it would release.[33]

1.3.2. Relations between Turkey and Iraq

Diplomatic relations between Turkey and Iraq are fashioned by three
principal factors:

a. the sharing of Euphrates river water;

b. security issues—tensions with the Kurdish minority in northern
   Iraq and southeastern Turkey;
c. close commercial links including a safe, continental conduit for
   Iraqi oil.

Turkey is a member of the North Atlantic Treaty Organization
and has positioned itself in international politics as a firm ally of the
United States since the days of the cold war. As such, Turkey offered
extensive bases and facilities to the United States and to its allies during
the Gulf War against Iraq's invasion of Kuwait. Despite this, Iraq does
not treat Turkey as an enemy, possibly because it simply has too much
to lose by doing so. Iraq often ignores Turkey's close alliance with its
enemies and maintains normal diplomatic relations with Turkey. Ministers from the two countries regularly pay visits to each other.

There are great economic interests at stake: Turkish-Iraqi relations
made very rapid progress from the mid-1970s until the end of the 1980s
on both economic and political fronts and after April 1982, Turkey
became the major outlet for Iraqi oil. In 1977, a first pipeline was laid
from Kirkuk in Iraq to Yumurtalik in Turkey and by the end of 1984,
the Kirkuk-Yumurtalik pipeline's capacity had been extended from 700,00
barrels per day to 1 million barrels per day. In 1985, Iraq and Turkey
began building a second pipeline through Turkey that was completed by
June 1987, increasing oil exports via Turkey from 1 million to 1.5
million barrels per day. Thus nearly half of Turkey's annual 20 million
tons of oil imports comes from Iraq (60 percent of Iraq's total output)
as well as $280 million in royalties per year for the oil transported via
these pipelines. The Turks often warned Iran not to attack these pipelines, as they were part of Turkey's vital economic zone.[34] Without the
Turkish outlet, Iraqi oil exports would have come to a virtual standstill
long ago.[35]

Indeed, Iraq and Turkey have another common interest: the sup-
pression of Kurdish dissidents in their frontier areas. This is the major
factor that has led to cooperation between the two countries. Both
States have been generally careful not to harbor or to support the other
side's opposition groups and when the Iraqi army was preoccupied with
fighting Iran in the Gulf War, Iraq even granted the Turkish army the
right of "hot pursuit" of Kurds across the border. After the closure of
the port of Basra during the Gulf War, road and rail links across Anatolia
and from Iskanderun became Iraq's back door for supplies. What is
more, new and enlarged pipelines carried Iraqi oil to the export termi-
nal in Yumurtalik near Mersin. Thus Iraq, heavily dependent on Turkey's
goodwill, cannot strongly protest about that country's unilateral usur-
pation of the Euphrates water.[36]

Turkey and Iraq have, however, faced several major problems over the last few years, particularly Baghdad's increasing concern over Turkey's GAP project and the feeling that it was left out of the Damascus-Ankara agreement of July 1987 over the sharing of the Euphrates water during the building phase of the Ataturk Dam. During his Baghdad visit in April 1988, Ozal declared that the Turkish-Syrian agreement was a temporary one and that the real treaty would be reached through tripartite talks to be held by the three countries.[37]

Following Iraq's invasion of Kuwait on August 2, 1990 and Turkey's compliance with the subsequent embargo resolution of the UN, the pipelines were closed.[38] The closing of the pipelines introduced a problem of distrust in the mutual relationship that will not be easy to overcome since, until recently, Turkey had conveyed the impression to Iraq that it viewed the pipeline as a commercial enterprise, protected from the uncertainties of politics. Turkey's closing of the pipelines has left the Iraqi government suspicious that the pipelines might be turned off whenever it suits Turkish economic or political interests.

After nearly a decade of UN sanctions against Iraq, Turkey is seeking to boost its trade relations with Baghdad, which have sunk drastically from a pre-war level of $2.5 billion annually. Turkish officials say the embargo has cost Turkey over $30 billion in lost trade. Limited exchanges between the two countries resumed when the oil-for-food program was introduced in 1996. Turkish trucks also smuggle diesel through the Habur border post on their way back from Iraq, but the United States have chosen to overlook this traffic that is seen as a form of compensation to Turkey. Before the Gulf War, Iraq was Turkey's fourth trading partner and main supplier of crude oil. Turkey currently earns only $600 million a year from trade and transit fees on the oil pipeline. Iraqi oil exports are pumped through Turkey, via a pipeline to the Mediterranean terminal of Ceyhan, but amounts are strictly restricted, as part of the $10.5 billion oil-for-food program that allows Baghdad to sell oil in exchange for food and medicine.

In 1997, Turkey and Iraq had signed a preliminary deal to build a gas pipeline with a annual capacity of 10 billion cu m, but restrictions on Iraq's exports of oil and gas have so far held up the project.

1.3.3. Relations between Syria and Iraq

Examining Syria and Iraq's complicated relationship, we can observe that although the issue of sharing the water of the Euphrates brought the two States to the brink of war in 1975, it appears that since then both States have avoided conflict over the water. Syria did not exploit its position as an upstream State to harm Iraq and, even more, the same subject caused Syria and Iraq to cooperate in 1990 against Turkey's development plans.

There are no two regimes more similar to each other, both structurally and ideologically than Syria and Iraq in the Arab world. Yet severe disagreements over political, diplomatic, economic, and strategic relations as well as personal rivalry between the leaders are what shapes the relations between the States.

From an economic point of view, Syria harmed Iraq by stopping the flow of oil in the pipeline from Kirkuk in northern Iraq to the ports of Baniyas and Tripoli on the Mediterranean Sea for the first time in December 1966 and then in March 1967. In 1972, Syria nationalized the Syrian section of the pipeline and in 1973 forced Iraq to double its transit payments. From 1976 to 1979, Iraq stopped the flow of oil in the pipeline in protest over the intervention of Syria in Lebanon and in 1982, Syria stopped the flow of oil in order to damage Iraq's war effort against Iran. Since then the pipeline has been closed.

Syria supported Iran in its war with Iraq in 1982 and then again, in 1990 it supported the alliance in its war to remove the Iraqi forces from Kuwait. This support by Syria in the war against Iraq did not only arise out of the severe rivalry between the leaders and the different ideological points of view of the Ba'ath parties, it also helped Syria to overcome a crisis connected with the end of the cold war and the collapse of the communist regime in the Eastern Bloc allowing it to begin to develop economic and political ties with the Western world.

In recent years, Syria and Iraq have normalized relations with each other and jointly criticized Turkey's construction on the GAP project. Both countries have been actively seeking support from the Arab League by demanding that Turkey consult with them over water rights. The Arab League issued several resolutions to this effect, claming that Turkey was allowing too little water to reach its neighbors and that the water coming from Turkey was polluted.[39]

In 2001, Syria and Iraq signed two key bilateral agreements intended to strengthen their alliance. On January 2001, the two States reached a new water-sharing agreement.[40] The details of the agreement have not been publicized. According to the BBC,[41] an Iraqi official claimed that Iraq and Syria have reached agreement on the use of the Euphrates and Tigris rivers and have called for Turkey to take part in their water-sharing negotiations. The Syrian-Iraqi accord covers a formula for sharing the waters of the Euphrates between the three countries. It is the first such agreement since 1990, when Syria promised to leave 58 percent of the Euphrates water to Iraq. Also on January 31,

Syria and Iraq signed a free-trade agreement reportedly worth $1 billion anually. The agreement includes the reopening (closed since the Iran-Iraq war in 1982) of a 552-mile pipeline linking the Kirkuk oil fields in northern Iraq with the Syrian port of Banias on the Mediterranean[42] and will likely be a huge boost to an Iraqi economy suffering from international trade sanctions.[43] Both Iraq and Syria have benefited from this reopening particularly as Syria is able to buy Iraqi oil at a cheaper than market price and to sell its own oil at market price. Syria's payments to Iraq bypass the contentious UN oil-for-food program.[44]

In January, Britain accused Syria of importing at least 100,000 barrels of oil per day in 2001 in breach of UN sanctions. Other estimates have placed the oil imports as high as 200,000 barrels per day, sending up to $1 billion to go directly to the Iraqi regime. Syrians widely view the sanctions regime against Iraq as an example of American double standards in the Middle East and do not believe that it should deter Damascus from forging a valuable economic partnership with Baghdad.[45] Bush administration officials say that they have repeatedly raised this issue with Damascus, demanding that the Iraqi exports, variously estimated at 150,000–200,000 b/d, must be brought in line with compliance with Security Council resolutions. So far, the Syrians have ignored these remonstrations, even though President Bush excluded Syria from the 'axis of evil'—Iran, Iraq, and North Korea—which he proclaimed on January 29, 2005.[46]

2. ANALYSIS OF RELEVANT INSTITUTIONS AND ACTORS

While generally it is an oversimplification to speak of States as homogeneous entities, it is actually fairly accurate in the case of the Euphrates riparians. Both Syria and Iraq have autocratic, authoritarian regimes and, while the Turkish government is representative parliamentary, the major water authorities—the water minister and the director general of the GAP project (a minister-level position)—are fairly autonomous and do not rotate with each government. Moreover there are no regional institutions to which the three belong—Turkey is a member of NATO, while Syria and Iraq belong to the Arab League; each has a different position on the 1997 Convention on Non-Navigational Uses of International Waters; and, with the exception of Turkey, there is little in the way of nongovernmental political activity. Therefore, in the case of the Euphrates, it is more correct to speak of interests, rather than institutions, which drive the relevant ministries—Agriculture, Water, and Foreign—to act both internally and externally.

2.1. Identification of the Relevant Interests

The conflict over the Euphrates water is a good example of a conflict ostensibly over freshwater which, on closer analysis, can be seen to be affected by many factors of which water is just one. The political economy of the catchment area, in which Euphrates water is a significant element, is influenced by the ideological, national, and security aspirations of the political entities that occupy the Euphrates Basin. In this conflict, it is not possible to determine which factor is foremost in shaping the approach of an individual riparian at any given time. In order to deal with such complexity, it is necessary to address the three major factors that appear to shape the conflict over water use by riparians namely economic interests, foreign political interests, and domestic policy (see Figure 8).

The major ideas and principles that underlie national interests and inspire the policy of national governments are discussed in the following sections.

2.1.1. Economic Interests

Economic factors are of great importance in setting priorities. Although water is nowhere treated as a pure economic good, economic theory does provide some guidelines for policy options regarding efficient water distribution. The three most important economic factors for developing the Euphrates Basin are food security, the production of hydroelectric power, and the development of industry. A riparian country's perception of future water needs determines whether it will be cooperative or uncooperative in its relations with other riparians concerning water. If a riparian nation foresees a great need for additional water in the future, it will be more defensive in negotiations with other countries. In contrast, if a riparian has little concern about water use, it will more readily agree to the institution of water allocation and management measures.[47]

Economic development, especially in its early stage, is often dependent upon agriculture. Agricultural demands for water always dominate water allocation in arid and semi-arid countries (80% in Turkey and about 77% in Iraq). However, the economic returns for agriculture, especially returns to water, are relatively low compared with those for industry. Industrialization has enjoyed a high priority in the planned government expenditure of all of the riparian States and when industry develops, the amount of water required increases.[48] Yet food security policies have the most important impact on water resources and are perceived by the governments of the riparian States as a matter of national priority. Food security, in turn, leads to special emphasis on agriculture that increases the demand for water and for "water security." Stress on water security leads to more tension with riparian neighbors over water, in turn leading to increased general concern over security and stronger defense of agriculture.[49] Nonagricultural uses of water, such as the production of hydroelectric power, will also have an impact on the water management of rivers through the evaporative losses from reservoirs. The availability of adequate power generation capacity is considered to be a prerequisite for rapid economic and social development and energy consumption has increased in all of the riparian countries very dramatically. However, the hydroelectric potential from the Euphrates River is very limited in Syria and Iraq because of the topographic conditions that do not allow for the construction of high dams.

2.1.2. Foreign Policy

Clearly, the decisions involved in developing policies for global river development are mainly political and can only be adequately addressed in political terms. It is recognized by international lawyers that international law can only be involved when treaties and mutually acceptable agreements have been signed and ratified. Global politics are among the most important features of international river basin development and have to be addressed in any analysis.[50] Several significant factors in international relations influence the degree and intensity of conflict over the Euphrates water. Factors such as national image, international law, and linkages with other issues such as territorial claims, border security, ideological conflict, and economic relations, can all influence a country's position in matters concerning water resource utilization. For example, Turkey claims that both Baghdad and Damascus wish to amplify the water issue to deflect attention from their other problems such as economy and internal instability.[51]

2.1.3. National Image

The concern for a positive national image may be one of the most important factors in deciding how to deal with international water issues. Turkey attaches great importance to the GAP project not only because of its economic and sociological benefits but also as a matter of national prestige. Syria also has a similar point of view toward the Tabqa Dam project and the economic projects linked to it. On the other hand, the autocratic regimes of Syria and Iraq do not like to be dependent on a non-Arab power, especially for the supply of water that is a very important resource, not only strategically and economically but also from a sociopsychological point of view.[52] Vulnerability to water scarcity contradicts the image of a serious and powerful leadership. The image of political weakness would become evident through palpable water shortages and could be politically destabilizing.

2.1.4. International Law

Another important component in international relations is the riparian attitude toward international law. Nations may either ignore or adhere to the principles of international law. Usually, riparians support only those principles of water law that favor their position in a basin, although they may accept widely espoused global principles even when they are to their disadvantage because they do not want to lose their credibility in the world community.[53]

The basic principles of international law in the matter of international rivers are based on equity and justice but do not provide any institutions that automatically establish operational procedures to implement the legal principles. "Equitable" does not mean equal use; rather it means that a large variety of factors, including population, geography, and the availability of alternative resources, can be considered in the allocation of water rights. The international law for international rivers is not mandatory and cannot compel States to solve conflicts. However, its principles are widely understood although not publicly accepted by riparian countries.

The development of nonbinding international legal principles can be an important factor in enabling a country to become involved in negotiations. Principles of global law could function as basic guidelines for the management of water equitably in the Euphrates Basin States. Turkey has drawn attention to the differences between "international" and "transboundary" water courses and has claimed privileges as a Euphrates upstream state over water sharing. Turkey also claims that according to international law, the waters of the Euphrates and the Tigris rivers have to be considered as one single basin and the sharing of the water among the riparian States should be related accordingly. On the other hand, Syria and Iraq consider the Euphrates River as international (common) waters and demand that they be shared according to "common use" or "according to their own needs." Syria's interests will relate to global law, particularly with reference to articles that deal with the "availability of alternative resources" because Syria is short of water resources compared with Turkey and Iraq. Iraq's claims on Euphrates water relate to the articles dealing with "prior uses" and "historical rights" and the articles dealing with the "population factor." It seems that the Turkish declaration of "absolute territorial integrity" has been made mostly for internal proposes but as long as Iraq and

Syria maintain their demands based on their lower riparian position, Turkey will insist on its advantage as an upper-riparian State.

The vote on the 1997 convention is telling: Syria voted yes, Turkey voted no, and Iraq did not cast any vote at all.

2.1.5. Ideological Conflict

The foreign relations of the riparians in the Euphrates Basin have fluctuated widely since they came under the influence of Western powers following World War I. After World War II, these Western powers attempted to bring the riparian countries into their coalition through military alliances such as Turkey's membership in NATO and later in the Baghdad Pact. From the 1960s, the Soviet Union began to play a more important role in the region and the political differences in the foreign relations of the riparians tended to reflect the conflicting interests of the superpowers in the area.[54] The collapse of communism in Eastern Europe and the breakup of Syria's leading ally, the USSR, at the end of the 1980s, left Syria without the important support of the Soviet Union. On the other hand, the Gulf War provided Damascus with an avenue for shifting its priorities to the Western camp. The end of the cold war and the Gulf crisis, however, have also changed Turkey's priorities and forced it to look further afield mainly toward the Caspian Sea and central Asian regions.[55]

2.2. Relationships and Interconnections between Issues

2.2.1. Linkage with Other International Issues

Linking conflicts over the water of a river basin with other bilateral or multi-lateral issues is one means that countries may use to extract concessions from their neighbors. Relations among the Euphrates riparian States are highly influenced by other factors such as territorial claims, border security arrangements, economic interests, and ideological conflicts. None of these factors relate directly to the water of the Euphrates River but they are on the agenda of many bilateral meetings and figure in other communications between political leaders. In fact, they play a very important role in bi-or trilateral negotiations among the riparian nations. For example, in 1990 President Saddam of Iraq demanded a formal sharing of the Euphrates water through a trilateral agreement before any economic agreement could be reached with Turkey.[56] In addition, Saddam's decision in 1990 to send his oil minister to Turkey concerning the water issue can be seen as a reminder of the wide range of concerns involved in maintaining good relations.

A number of issues adversely affect Turkish-Syrian relations, the most important is the Syrian support of the PKK that started immediately after Turkey's construction of the Ataturk Dam and its talk of the "stolen province of Alexandretta." Shortly after the Gulf crisis erupted, Syria adopted a pro-Western approach and, as a result, Turkish-Syrian relations improved. Another example is that Turkish actions concerning the Euphrates could be seen as political maneuvers carried out in association with U.S. interests to continue pressuring Syria through Lebanon.[57]

Iraq and Turkey have fewer linkage issues that threaten their relations as they have no territorial claims upon each other. Until Iraq's invasion of Kuwait in August 1990, they had considerably strong economic and political relations as well as common interests in their fight against the Kurds in northern Iraq. It is important to remember that their mutual economic relations were good before the Kuwait War and it is logical to assume that they will again be so in the future. All of these factors have enabled them to create and maintain good relations.

Iraq and Syria's relations have been heavily influenced by the ideological conflict between the separate Ba'ath regimes of the two countries, as well as by the personal hostility between the leaders, the closing of the pipeline connecting Iraq with the Mediterranean Sea, Syrian support of Iran during the Iran-Iraq War, and its coalition with the UN during the Iran-Iraq Gulf War.

2.2.2. Domestic Interests

There are several important factors that have to be taken into account when setting priorities and resources concerning the development plans of a country. For example, priorities have to be set vis-à-vis population dispersion in this case mainly because rural to urban migration accounts for about one-third to one-half of the growth of the main cities in Turkey and Iraq. It is also necessary to plan how to achieve improvements in the standard of living of the population in rural areas and in internal security (the Kurdish population in southeast Turkey and the Shia population in Iraq). The development of hydroelectric power, agriculture, and industry are aimed mainly to achieve these targets.

The political stability of the riparians also influences the nature of the conflict since a weak nation might be too preoccupied with its internal problems to become actively involved in external disputes. Conversely, a State beset with international difficulties may well choose to divert attention from its domestic difficulties and attempt to bring about a modicum or semblance of internal unity in the face of a presumed external danger.[58]

National emotions have also been mobilized with respect to water as well as the personal prestige of the riparian leaders and these also affect the setting of priorities concerning water development plans. Such is the case with the dams on the Euphrates River that have acquired a psychological importance as a symbol of development. The Euphrates project in Syria that has been given a high political priority is an example of a showcase of the Ba'ath development drive, especially in its naming the reservoir after President Assad.

In these circumstances, internal disputes have always been likely between entrenched rural agricultural interests fighting to retain an advantageous position vis-à-vis growing urban and industrial interests. However, if a government fails to secure food supplies, whether from internal or external sources, an atmosphere of political disquiet will emerge. Thus a natural political alliance has arisen between the countries' political leaderships and their rural communities since the former want to ensure national security and the latter want to provide food for national needs.[59] The political power of agriculturists has enabled the agricultural sector to exert powerful pressure on national water policy and policy-makers respond to these pressures in the development of their national and international policies.

2.3. Qualification of These Links

One approach that assists in the analysis of the factors considered by each riparian is that of "cognitive mapping." With this approach it is possible to obtain a type of mental map or cognitive structure of issues as seen by the relevant actor.[60] A comparison made between the cognitive maps of an issue held by each major actor is helpful in locating critical differences that may determine policy and behavior, as well as for ascertaining whether the issues are real. The technique enables the analyst to identify the factors involved in the network of relationships between the States as well as to determine whether perceptions are similar and whether problems lie in conflicting interests. It also indicates whether a significant part of the conflict arises from the utilization of discrepant cognitive maps that might be harmonized through information, negotiation, or by other means. The mapping permits a basic analysis of power. A full application of this technique allows for the mapping of all of the riparians in the Euphrates Basin, the identification of critical subnational actors, and the relationship of decision-makers to the complexity of the issues as well as to their perception of the policy environment.

The cognitive mapping used here will examine the national interests of the three riparian states during three periods of time:

    a. before the construction of the major dams, when natural factors in large measure determined the availability of water to the three riparians;

b. the period between the 1960s and 1990s that was characterized by rapid development in all of the riparian countries; and

c. the future period, in which the riparian development plans will be completed.

The thickness of the lines in Figures 9–11 indicates the perceived strength of association between factors and interests: the thin line represents low-level interests; the medium thick line the high-level interests; and the thickest line, the interests that are critical from the point of view of the individual States. The importance of the various factors in the determination of the level of the States' interests in the river water is based on the data accumulated in the previous chapters. A crucial error in some cognitive maps is the omission of the overall importance (the evaluation or utility) of the cognitive realm being mapped. In the present instance, however, since "natural interests" is a main component and is obviously of utmost importance, this limitation is not significant.[61] The more connected or embedded a cognitive element is, the more significant it is and the more resistant it is to alteration or replacement. More details on the links and projections are found in the following sections.

3. MANAGEMENT OF THE RIVER BASIN

3.1. Assessment of the Actual Management of the River Basin

The three cognitive maps clearly reflect the level of the riparian States' interests in the waters of the Euphrates River over three periods of time: the past, before the implementation of the large development projects for the river water and when the use of water was minimal; the present, from the 1960s to the 1990s when the three States were at the peak of their implementation of development plans and the resultant use of the river water grew; and the future, after the completion of the development projects when the resultant use of the river water reaches its peak. A number of very significant wide-ranging conclusions can be drawn from the three figures.

3.1.1. The period up to the 1960s was characterized by scanty use of the river water, mainly by Syria and Turkey that are located upstream. The natural flow of the river actually reached Iraq that invested most of its efforts during this period in projects aimed at preventing flooding. Syria and Turkey, in contrast, were planning the construction of dams along the river and Turkey, in particular, was planning dams for the production of electricity that would only marginally affect the amount of water in the river. Syria was planning the construction of a multi-purpose dam for the production of hydroelectric energy and for irrigation purposes whose potential effect on the flow of water to Iraq was very limited.

Since there were, as yet, no development projects for the river in Turkey and Syria, the level of economic expectations and internal interests of the States was particularly low. As a result, the effect of river water usage on the foreign relations between the States was limited.

This period can be summed up as one with a limited range of low-interest level factors of marginal importance in their effect upon the network of relations between the riparians, with a limited demand for water and a low potential for conflict over the river water as well.

3.1.2. The period between the 1960s and the 1990s was characterized by the accelerated implementation of development projects by the three riparians that was expressed in the building of dams for the production of hydroelectric power and dams that combined the production of electricity with wide-ranging irrigation projects. As a result, the level of States' interests significantly increased and additional factors of interest accumulated.

From the economic point of view, Turkey's three dams were expected to produce hydroelectric power that would have an important effect upon the oil-poor country. Turkey had not yet begun to exploit the river water for the irrigation of large areas in southeast Anatolia.

Syria intended to produce hydroelectric power with the Tabqa dam and this raised high expectations although they were not to be satisfied. Irrigation projects that demanded growing amounts of water were a central goal for Syria although is was already clear that both the area irrigated and that expected to be irrigated would be smaller than hoped for.

In contrast to the two upstream States, Iraq neglected the development of the river which, as far it was concerned, was unimportant as potential for hydroelectric power and preferred to base its economics upon the importation of feed products over the development of its agriculture.

Economic development projects reflect the development of many other factors and the domestic interests of the States that were of marginal interest during this period now became very significant. Such development was significant for Turkey that was trying to prevent the negative migration from the area by raising the standard of living. It was further expressed by an attempt to improve the shaky internal security that had arisen because of the Kurdish revolt. The prestige of Turkey's leadership was also significantly linked to the success of the development project.

Syria encourages migration to the Euphrates River area (in order to reduce the pressure on the urban networks systems) by raising the standard of living in those areas. The development of the Euphrates Valley became a matter of personal interest for the Ba'ath Party and for President Assad.

These development projects had major significance for the network of foreign relations among the riparians that itself was directly connected with the international power system (Turkey, a non-Arab state, a member of NATO, and with a Western form of government; Syria and Iraq mainly influenced by the foreign policies of the USSR; and with autocratic, authoritarian governments).

The system of dams built in Turkey, which allowed it to control the flow of the river water, changed the balance of power in the region. Syria and Iraq made considerable efforts to reduce the damage involved as much as possible. Iraq and Syria that were on the brink of war in 1974, here cooperated against their common source of danger, Turkey. In the first stage they tried to prevent and later to hold up, the building of the Ataturk Dam by trying to delay granting international financial assistance for constructing the dam. Syria began to support the Kurdish rebels during this period and Iraq complicated its economic relations with Turkey. This period symbolizes one of significant tension in foreign relations between the States, since there was always the potential for instability linked to the ever-growing demand for water and change in the strategic balance.

Management considerations for each State are listed in the following section, divided by foreign relations, economics, and domestic interests.

3.1.1. Turkey

Turkey's domestic interests up to the 1960s were seen as a factor with high implications for the level of interests in the river water as opposed to foreign relations that was seen as being only secondary. From the 1960s on into the future, economic interests and foreign relations are perceived as having much more significance than domestic interests.

a) *Foreign Relations*

National image and political influence up to the 1960s were more significant as opposed to factors of ideological conflict and border security that were of no significance during this period. In contrast, during the 1960–1990 period the factor of political influence appeared and territorial claim joined the list of insignificant factors. In the future, ideological conflict will also lose its importance.

b) *Economics*

Until the 1960s, the economic interests of the Euphrates River were perceived as insignificant. From the 1960s on into the future, the production of electricity is perceived as a factor that produces the most significant level of interests, followed by agriculture, while industry and services were found to have only marginal importance.

c) *Domestic Interests*

Until the 1960s, the standard of living and internal security were perceived as being most significant as opposed to population distribution and food security. Between 1960 and 1990, population distribution was of paramount importance as opposed to the standard of living and the image of Turkish leaders. For the future, the standard of living, domestic use, and internal security are perceived as being more important than the image of Turkish leaders and the goal of population distribution.

3.1.2. Syria

Until the 1960s, domestic interests were considered the most significant of Syria's interests in the river water as opposed to foreign policy. In contrast, from the 1960s, foreign policy and the economy are seen as most significant as opposed to domestic interests.

a) *Foreign policy*

National image and territorial claim were perceived as more important than ideological conflict, international law, and economic relations. During the 1960–1990 period, territorial claim, national image, and political influence were evaluated as more important than international law, ideological conflict, and border security. In the future, economic relations and international law will become more significant and border security and ideological conflict will decrease in importance.

b) *Economy*

Until the 1960s, only agriculture had a marginal level of importance. From this period on, production of hydroelectric power as well as agriculture became important and only industry and services was found to be marginally important.

c) *Domestic Interests*

Until the 1960s, domestic use was perceived as most significant as opposed to the standard of living, population distribution, and food security. The standard of living, the image of the leaders, and population distribution became most significant from the 1960s on into the future in contrast to internal security.

3.1.3. Iraq

Until the 1960s, domestic interests were found to be most significant for Iraq and foreign policy had secondary importance. In the 1960–1990 period, the most important interests were the economy as opposed to domestic interests. In the future, domestic interests are expected to again become most significant, not foreign policy.

a) *Foreign Policy*

Political influence, border security, and national image were most important during the 1960s as opposed to ideological conflict and international law. During the 1960–1990 period, political influence, national image, and economic relations were the most significant. International law was considered marginal up to the 1990s but will, in the future, become a major factor as opposed to international security.

b) *Economy*

Grading the economic factors from the past into the future, the experts rank agriculture as the most significant factor and the production of hydroelectric power and industry and services as having only marginal importance. Domestic interests, such as internal security, were perceived as most significant from the past to the 1990 as opposed to population distribution, standard of living, and food security. The factors of internal security and the image of the leaders will, in the future, decrease in importance in contrast to food security and preserving the standard of living.

3.2. *How the Situation Is Likely to Evolve*

3.2.1. Turkey

When the development plans have been implemented, Turkey's *economic expectations* will likely increase, based on the Euphrates water and its dependence on them. Meanwhile Turkey is, for all practical purposes, self-sufficient in food and it wishes to remain so in the future. The present agricultural production in the GAP area is limited to one harvest every two years but with irrigation and modern technology, the soil and climate could sustain two or three harvests annually.[62] Turkey, however, expects to profit from the export of crops grown in the irrigated areas of GAP because southeast Turkey is located in a favorable geographic position to deliver fresh and packaged foods to both the Middle Eastern countries and Europe, cheaper and in better condition than any other supplier. Markets have also been identified in neighboring countries and Turkey plans to export agricultural products to Iraq and Syria, both of which face food deficits (Turkish exports to Iraq in 1988 were $986 million, and exports to Syria were $110 million).[63] When the GAP project is completed, it is estimated that Turkey will produce enough food to feed 80 million people and 3.3 million extra jobs will have been created countrywide. The increase in agricultural production may also be expected to start a "chain reaction" in other sectors of the regional economy; thus Turkey perceives that it cannot tolerate any limitation on its economy caused by any delays in the GAP project.[64]

The extensive irrigated areas will contribute to the creation of jobs, raise the standard of living, become a source of foreign currency resulting from the export of foodstuffs, and provide an infrastructure for the further industrial development of the region. Turkey thus hopes that it will put a stop to the negative emigration from the region and that the problems of internal security will diminish.

In the area of *foreign policy,* it seems that the problem of the security of Turkey's frontiers will abate after the signing of a security agreement with Syria, although some military tension is still to be expected on the Iraqi border involving the Kurdish opposition. There is no forecast that the subject of the territorial demands made by Syria and Turkey will significantly affect the foreign relations of the countries, since nowadays the end of the cold war makes it possible for the normalization of their mutual relations to take effect. The subject of economic relations will assume much more significant dimensions in the future and there is no doubt that the decision about whether Turkey will become a member of the European Community or not will have implications for the importance of the economic ties between Turkey and the neighboring countries. Turkey is also a member of NATO and wishes to play a certain role in that organization, has important trade and financial relations with Middle East countries, and is interested in economic opportunities in North America, the Black Sea area, and central Asia. Hence, its international image is a rather important concern.[65]

In the area of *internal policy,* the subject of the redistribution of the population will be of cardinal significance but as the standard of living of the population of southeast Anatolia rises, this source of instability will lose some of its pivotal importance. The extensive use of the Euphrates water for the development of irrigation and industry will promote a rise in the standard of living. Turkey looks forward to the widespread development of the municipal systems in the region. The river water use is intended to provide for the municipal and rural water needs that will greatly increase in the future.

3.2.2. Syria

It seems that the *economic importance* of agriculture based on river water will increase as a result of Syria's rapid demographic growth and its aspiration to produce enough food to meet its population's food requirements. However, the importance of river water, to Syria's economy will decrease through the generation of hydroelectric power because the potential to generate electricity is restricted and because the growth of the Syrian economy will eventually rest on generating thermal electricity. Therefore, the production of electricity on the Euphrates as part of the Gross National Product (GNP) will gradually decrease.

In *foreign policy*, the issue of the security of the frontiers between Syria and Turkey will gradually lose some of its significance as a result of the agreements made between the two countries. The ideological conflict will dwindle as one of the consequences of the crumbling of the Soviet Union. These new circumstances will also contribute to the development of the mutual economic interests of the countries.

The political tension between Syria and Iraq will continue on an ideological basis, at least as long as Saddam remains the ruler of Iraq. The same may be said of the tension involved in the distribution of river water that turns Iraq into the principal casualty of the agricultural development of the upstream regions, as regards both the quantity and the quality of the water.

As to *internal policy*, the redistribution of the population will continue to be a central issue in Syria that aspires to transfer two million of its inhabitants to settlements in the Jezirah region. Agricultural and industrial development are aimed at providing the inhabitants with a standard of living that will persuade them to move into the region. Consequently, more water will gradually serve domestic and industrial needs. Unlike Turkey, Syria has no internal security problems in the region nor any emigration of inhabitants moving to the country's big cities from the region.

3.2.3. Iraq

The principal importance of Euphrates water to Iraq, from an *economic point* of view, is linked to the development of agriculture. The share of hydroelectric power (out of all the power produced) will gradually decrease even further. A large number of the projects developed in Iraq, whose principal aim was to prevent flooding, will become redundant as a result of the constant flow of the river caused by the controlled release of water from the dams in Turkey and Syria. Conversely, the quantity of water in the river will be reduced and its quality impaired. This may affect the extent of agricultural areas and the type of production but Iraq has an alternative option, the use of the Tigris River water, transferred to the Euphrates through Lake Tharthar and the construction of a third river that will add agricultural land to the country.

In terms of *foreign policy*, Iraq's dependence upon Turkey and Syria will grow because the upstream States will determine the quality of water flowing in the river. The close economic ties with Turkey may be expected to be renewed and even strengthened after the embargo is lifted. As for Iraq's relations with Syria, only a change in the latter's regime will improve the fabric of foreign and economic relations between the countries.

With regards to *internal policy*, the Euphrates waters will continue to be important if Iraq wants to raise the standard of living of the local population and thus prevent negative emigration from rural areas. To achieve this, more and more water will be diverted to domestic use.

The Euphrates water is also associated with the image of the country's leaders who wish to see massive development in the region in order to restrict the extensive importation of food into the country.

3.3 What Should Be Done?

The Euphrates River system represents a complex set of water-related issues involving the geographic conditions of the area, the external and internal policies, and the different economic approaches of the countries of the drainage basin. An analysis of the factors that influence the system of relations between the States shows that the region is very heterogeneous. Factors that contribute to interstate conflict, like those that contribute to interstate cooperation, are influenced by geographic conditions, internal and external policies, economic development, ethnic, cultural, and ideological differences, as well as the personal relations between the leaders.

It is important to note that the countries are carrying out their development programs with no overall agreement for the division of water. The programs are similar in their aims: control of the river water; the production of hydroelectric power; the irrigation of arid and semiarid regions; and industrial development. However, due to the different geographic, political, and economic conditions, the States operate according to different agendas and motivations. Thus tension over the river water will probably continue to be a central issue in the framework of the nations' foreign relations.

The findings of this study can be summarized by a number of additional conclusions for the future.

1. It is impossible to define the Euphrates system as a closed system especially in the case of Iraq where other resources like the Tigris River flowing through the Tharthar/Euphrates canal also play a role in satisfying the demand for water. Thus the overall water balance of the Euphrates, the Tigris, and their tributaries needs to be examined in order to determine the water needs of the riparian States and to be taken into consideration regarding the sharing of the water.

2. There is sufficient water in the Euphrates system to meet the social and economic needs of the riparians until 2025, notwithstanding the proposed projects in Turkey. This adequacy relates also to the capacity of Syria and Iraq to utilize water that has been, and will continue to be, restricted by environmental and institutional impediments. Thus it appears that the division of the water will not be a cause for war between the States and that other political and economic factors play a decisive role in determining the network of relations between the States. Neither Syria nor Iraq will lack the water necessary for existing developmental projects in the near future but this fact does not solve the problem of their inability to ensure their food security. Therefore, the solution to any periodic and long-term water deficits will be in the global political economy in which the riparians have already participated and will continue to participate according to their individual interests as opposed to their sometimes conflicting interests. The political economy of water is subordinate to the global political economy of food and to the political economy of trade in food.

3. Since there is no cooperation between the riparians for the common development of the river basin, the individual economies can and will solve their problems independent of their riparian neighbors in the global system which, in the future, will lead to closer economic and political cooperation between the States.

4. Principles of international law are unlikely, in the case of the Euphrates, to be a significant basis for a procedure for sharing water and since the principles and instruments of international law are difficult to put into practice, the riparians will have recourse to other remedies that are applicable such as international trade.

5. Water sharing dominates the current thinking of the downstream riparians but they will increasingly adapt to the idea that they can find substitutes for water as they recognize the fact that this

solution has been the remedy to their food production problems during the past two decades and will remain a major remedy in the future.

How might negotiations proceed? Oregon State University houses The Transboundary Freshwater Dispute Database,[66] which currently includes a computer database of 150 water-related treaties and 39 U.S. interstate compacts, cataloged by basin, countries, or States involved, date signed, treaty topics, allocation measures, conflict resolution mechanisms, and nonwater linkages. The database also includes a digitized inventory of international watersheds, negotiating notes, and background material on 14 case studies of conflict resolution, news files on cases of acute water-related conflict, and assessments of indigenous/traditional methods of water conflict resolution.

What one notices in the global record of water negotiations is that many of the negotiations surveyed begin where the Euphrates is now, i.e., with parties basing their initial positions in terms of rights—the sense that a riparian is entitled to a certain allocation based on hydrography or chronology of use. Upstream riparians often invoke some variation of the Harmon Doctrine, claiming that water rights originate where the water falls. India claimed absolute sovereignty in the early phases of negotiations over the Indus Waters Treaty, as did France in the Lac Lanoux case, and Palestine over the West Bank aquifer. Downstream riparians often claim absolute river integrity, claiming rights to an undisturbed system or, if on an exotic stream, historic rights based on their history of use. Spain insisted on absolute sovereignty regarding the Lac Lanoux project, while Egypt claimed historic rights against first Sudan, and later Ethiopia, on the Nile.

In almost all of the disputes that have been resolved, however, particularly on arid or exotic streams, the paradigms used for negotiations have not been 'rights-based' at all—neither on relative hydrography nor specifically on chronology of use, but rather 'needs-based.' 'Needs' are defined by irrigable land, by population, or by the requirements of a specific project. In agreements between Egypt and Sudan signed in 1929 and in 1959, for example, allocations were arrived at on the basis of local needs, primarily of agriculture.

Once negotiations move from rights to needs, it is also useful to think in terms of "baskets of benefits." In most treaties, water issues are dealt with alone, separate from any other political or resource issues between countries—water qua water. By separating the two realms of "high" and "low" politics, or by ignoring other resources that might be included in an agreement, some have argued, the process is either likely to fail, as in the case of the 1955 Johnston accords on the Jordan, or more often to achieve a suboptimum development arrangement, as is currently the case on the Indus agreement, signed in 1960. Increasingly, however, linkages are being made between water and politics, and between water and other resources. These multi-resource linkages may offer more opportunities for creative solutions to be generated, allowing for greater economic efficiency through a "basket" of benefits. Some resources that have been included in water negotiations include financial resources, energy resources, political linkages, transportation infrastructure, and data.[67]

Finally, what would a basin-wide institution look like? Despite the tendency of water managers to think in terms of total integration of watersheds, the Euphrates is not the most likely setting for this to occur. Even friendly States often have difficulty relinquishing sovereignty to a supralegal authority, and the obstacles only increase along with the level of suspicion and rancor.

At best, one might strive for coordination over integration. Once the appropriate benefits are negotiated, it then becomes an issue of "simply" agreeing on a set quantity, quality, and timing of water resources that will cross each border. Coordination, when done correctly, can offer the same benefits as integration, and be far superior to unilateral development, but does not threaten the one issue all States hold dear—their very sovereignty.

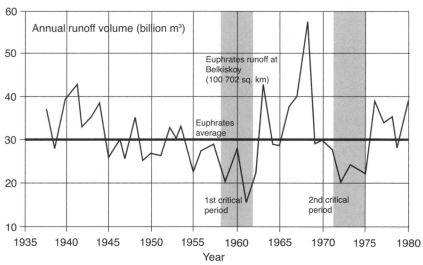

Source: Bagis, 1939, 34.

FIGURE 6. Euphrates River: historical annual runoff at Belkiskoy (1937–1980)

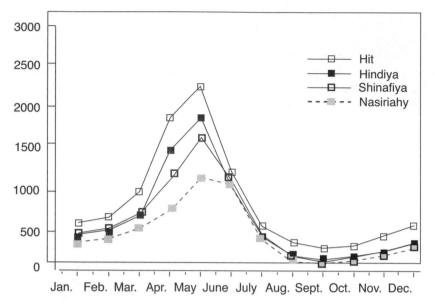

Source: Haige, 1951, 3.2

FIGURE 7. The Euphrates River mean monthly discharge in Iraq, 1926–1948

TABLE 9. Hydrological data for the Euphrates River

	Turkey	Syria	Iraq	Saudi Arabia	Total
Drainage area					
(1,000 km²)	125	76	177	66	444
% of total area	28	17	40	15	100
% of country	16	41	39	5	

Annual discharge	Turkey	Syria		Iraq	
(10 /m³)	OUTFLOW	INFLOW	OUTFLOW	INFLOW	
(1937–1963)					
Minimum	12,600	12,600	14,000	14,000	
Mean	28,400	28,400	32,400	32,400	
Maximum	42,000	42,000	45,000	45,000	
% of mean discharge	88		12	0	

Source: Beaumont, 1978, 37.

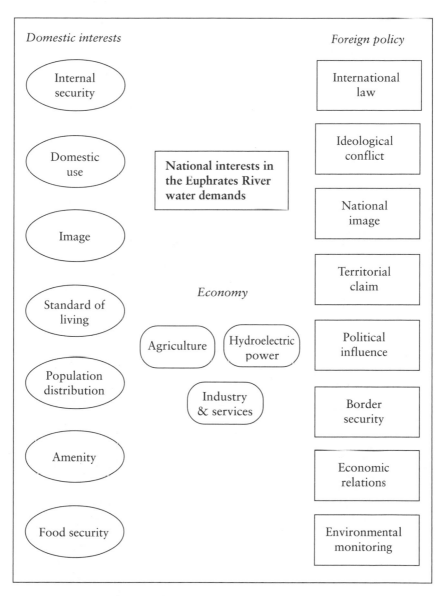

FIGURE 8. Factors where Euphrates water has an impact on domestic policy and international affairs

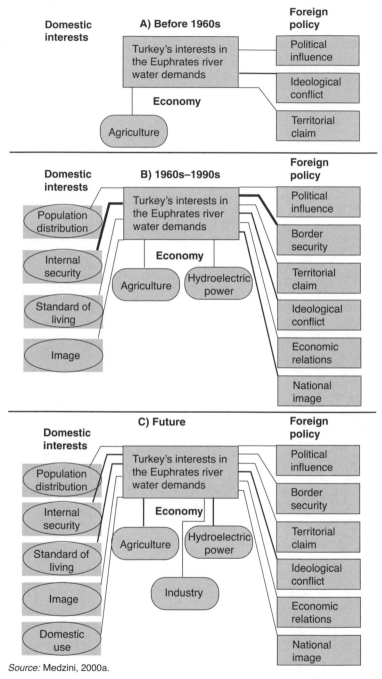

Source: Medzini, 2000a.

FIGURE 9. Cognitive maps of Turkey's interests in the Euphrates River: Pre-1960s, 1960s–1990s, future

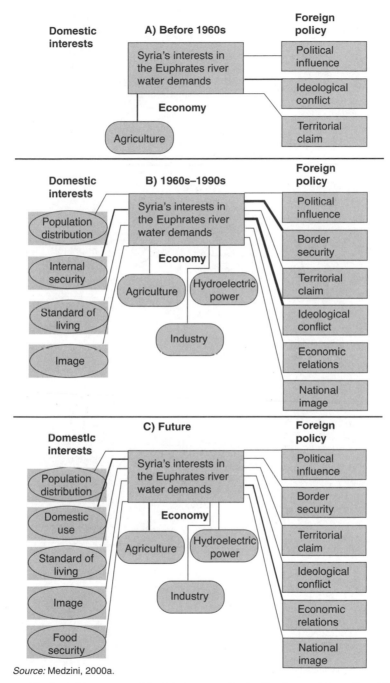

FIGURE 10. Cognitive maps of Syria's interests in the Euphrates River: Pre-1960s, 1960s–1990s, future

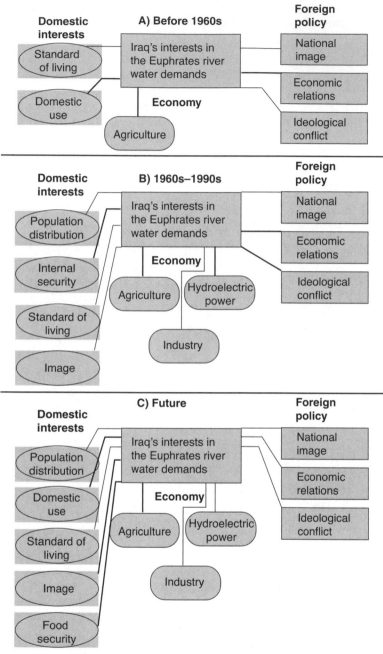

Source: Medzini, 2000a.

FIGURE 11. Cognitive maps of Iraq's interests in the Euphrates River: Pre-1960s, 1960s–1990s, future

NOTES

1. Al-Hadithi, 1979, 45.
2. Al-Khashab, 1958, 8.
3. Naff & Matson, 1984, 82.
4. Abbas, 1984, 69.
5. Beaumont, 1978, 35.
6. Abbas, 1984, 79.
7. GAP, 1990, Vol. 4, E6.
8. Qudain, 1960, 57
9. Ibid.
10. Hole & McCorriston, as reported in the *New York Times*, 2 April 1991.
11. El-Yussif, 1983, 19–22.
12. Wolf, 1998, 251–265.
13. Beaumont, 1991, 1.
14. Lowi, 1993, 108.
15. Naff & Matson, 1984, 94.
16. Kolars & Mitchell, 1991.
17. Naff & Matson, 1984.
18. Kolars & Mitchell, 1991.
19. Ibid.
20. Gruen, 1993.
21. Cited in ibid.
22. Turkish Embassy, Washington DC, June 1996.
23. MEI, 16 February 1990, 13; Bolukbasi, 1990, 3.
24. Middle East, October 1987, 27.
25. Bolukbasi, 1990, 24, 41.
26. Ibid., 34.
27. Lewis, 1991, 73.
28. Bolukbasi, 1990, 43.
29. Briefing, 20 July, 1987, 15.
30. CNN.COM, 11.20.2000.
31. CNN.COM, 03.11.2002.
32. Lovatt, 2002, 7.
33. Lupo, 2002, 538.
34. Barkey, 2000, 111.
35. Inan, 1989, 51; Bolukbasi, 1990, 22.
36. MEI, 16 February 1990, 13.
37. Bolukbasi, 1990, 39.
38. Lewis, 1991, 69.
39. Lupo, 2002, 537.
40. ArabicNews.Com, 1/29/2001.
41. BBC News, 1.31 2001.
42. ArabicNews.Com, 1/25/2001.
43. Lupo, 2002, 358.
44. Lovatt, 2002, 9.
45. Blanford, 2002, 6.
46. Middle East, June 2002, 23.

47. Abbas, 1984, 208.
48. Tuma, 1987, 131.
49. Frey, 1993, 63.
50. Rogers, 1991, 14.
51. Tekeli, 1990, 212.
52. Ergil, 1991, 52.
53. Abbas, 1984, 210.
54. Ibid., 225.
55. Briefing, 18 January 1993, 5; MEI, 16 April 1993, 16.
56. Tekeli, 1990, 210.
57. Nasrallah, 1990, 16.
58. Abbas, 1984, 207.
59. Allan, 1992, 5.
60. Frey & Naff, 1985, 74.
61. Naff & Matson, 1984, 185.
62. Tekeli, 1990, 208.
63. Ergil, 1990, 11.
64. Inan, 1990, 5.
65. Frey, 1993, 65.
66. <http://terra.geo.orst.edu/users/tfdd/>.
67. See Wolf 1998, for more information.

BIBLIOGRAPHY

Abbas, H. A. H. (1984). *Managing Water Resources in the Tigris and Euphrates Drainage Basin: An Inquiry into the Policy Process.* Ph.D. diss., North Texas State University.

Al-Hadithi, A. H. (1979). *Optimal Utilization of the Water Resources of the Euphrates River in Iraq.* Ph.D. diss., University of Arizona Graduate College.

Al-Khashab, W. H. (1958). *The Water Budget of the Tigris and Euphrates Basin.* Ph.D. diss., University of Chicago.

Allan, J. A. (1992). *"Striking the Right "Price" for Water: Achieving Harmony between Basic Human Need, Available Resources and Commercial Viability,"* Paper given at the Conference on Middle-East Water Resources convened at the School of Oriental and African Studies by SOAS and IBC, 19–20 November.

Anonymous (1999). "New Moves on Jordan Valley Projects" *MEED Middle East Economic Digest.* 43:(1), p. 20.

ArabicNews.Com, 7/20/2000. "Syrian Efforts to Rationalize Water Consumption." http://www.arabicnews.com/ansub/Daily/Day/000720/2000072013.html.

ArabicNews.Com, 1/25/2001. "Iraqi official and the Banias-Karkouk Oil Pipeline." http://www.arabicnews.com/ansub/Daily/Day/010125/2001012514.html.

ArabicNews.Com, 1/29/2001. "Syrian-Iraqi Talks on Waters, Housing Issue." http://www.arabicnews.com/ansub/Daily/Day/010129/2001012916.html.

Bakour, Y. and J. Kolars (1994). The Arab Mashrek: Hydrologic History, Problems and Perspectives. In *Water in the Arab World: Perspectives and Prognoses*. Rogers, P. & Peter Lydon eds. Division of Applied Sciences, Harvard University.

Barkey, H. J. (2000). "Hammed in by Circumstances: Turkey and Iraq since the Gulf War," *Middle East Policy*: (4), October 2000, pp. 110–126.

BBC NEWS, 26.02. 2000. "Iraq and Syria Resume Ties." http://news.bbc.co.uk/1/hi/world/middle_east/658371.stm.

BBC NEWS, 3. 01.2001. "Syria and Iraq to Boost Trade." http://news.bbc.co.uk/1/hi/world/middle_east/1146804.stm.

BBC NEWS, 31.01.2001. "Syria and Iraq Sign Trade Deal." http://news.bbc.co.uk/1/hi/world/middle_east/1141960.stm.

BBC NEWS, 28.01.2001. "Syria and Iraq Hold Water Talks." http://news.bbc.co.uk/1/low/world/middle_east/1142000.stm.

Beaumont, P. (1978). "The Euphrates River—An International Problem of Water Resources Development." *Environmental Conservation 5*, pp. 35–43.

Beaumont, P. (1991). "Transboundary Water Disputes in the Middle East." Submitted at a conference on Transboundary Waters in the Middle East, Ankara, Turkey, September 1991, p. 1.

Beaumont, P. (2000). Conflict, Coexistence, and Cooperation: A Study of Water Use in the Jordan Basin. In *Water in the Middle East*, Hussein A. and A. Wolf eds. Austin: University of Texas Press.

Biswas, A., J. Kolars, M. Murakami, J. Waterbury, and A. Wolf, eds. (1997). *Core and Periphery: A Comprehensive Approach to Middle Eastern Water*. Water Management Series 5. Delhi: Oxford University Press.

Bleier, R. (1997). "Will Nile Water Go to Israel? North Sinai Pipelines and the Politics of Scarcity." *Middle East Policy, 5*:(3), p.113.

Bolukbasi, S. (1990). *Turkey Policies Challenged by Iraq and Syria: The Euphrates Dispute and the Kurdish Question* (unpublished manuscript, Istanbul).

Briefing (1987). "The PKK Threat: a Myth or a Reality." *Briefing*, 20 July pp. 8–12.

Briefing (1993). "Demirel to Damascus for 'Watershed' Visit." *Briefing*, Issue 923, 18 January pp. 4–5.

Campagna, J. (1999). "The Politics of Water (Israel Cuts Water Supplied to Jordan)." *World Press Review*, 46 (8) p. 20.

CNN.COM, 11.20.2000. "The Kurdish Question after Ocalan." http://www.cnn.com/2000/WORLD/europe/11/20/kurdish.question/index.html.

CNN.COM, 08.03.2002. "Turkey Abolishes Death Penalty." http://www.cnn.com/2002/WORLD/meast/08/03/turkey.death.pen/index.html.

El-Yussif, F. (1983). "Condensed History of Water Resources Developments in Mesopotamia." *Water International*. 8, pp. 19–22.

Ergil, D. (1990). "Politics of Water." *Siyasal Bilgiler Fakultesi Dergisi,* Ankara Univrsitesi, nos. 1–4, December, pp. 53–80 (in Turkish).

Ergil, D. (1991). "The Water of Turkey and International Problems." *Dis Politika Bulteni,* no. 1, April, pp. 40–62 (in Turkish).

Frey, F. W. (1993). "The Political Context of Conflict and Cooperation Over International River Basin," *Water International*, 18, pp. 54–68.

Frey, F. W. and Naff, T. (1985). "Water: An Emerging Issue in the Middle East, Annals, *AAPSS*, no. 482, November, pp. 65–84.

GAP (1990). *The Southeastern Anatolia Project Master Plan Study,* final master plan report, Vols. 1–4, June, Republic of Turkey Prime Ministry State Planning Organization.

Gruen, G. (1993). "Recent Negotiations over the Waters of the Euphrates and Tigris." Proceedings of the International Symposium on Water Resources in the Middle East: Policy and Institutional Aspects. Urbana, IL, 24–27 October.

Hof, F. (2000). Water Dimensions of Golan Heights Negotiations. In: *Water in the Middle East*, Hussein, A. and A. Wolf eds. Austin: University of Texas Press.

Inan, K. (1989). "The Southeastern Anatolia Project and its Contribution in the Middle East." In *Studies on Turkish-Arab Relations*. Soysal, E., ed., Annual 4, pp. 48–60.

Inan, K. (1990). "Southeastern Anatolia Project and Turkey's Relations with the Middle-Eastern Countries." *Middle East Business and Banking*, 9: (3), March, pp. 4–6.

Inan, Y. (1992). Department of International Relations, Faculty of Economics and Administrative Sciences, Gazi Universitesi, Ankara, *Interview*, 8 July.

Israel Ministry of Foreign Affairs (2000a). *The Peace Process Reference Documents,* http://www.israel.org/mfa/go.asp?MFAH00p0.

Israel Ministry of Foreign Affairs (2000b). *Israel's Environmental Laws* gopher://israelinfo.gov.il:70/00/govmin/envir/envlaw/frwater/941032.evl.

Israel Ministry of Foreign Affairs (2000). *Selected Laws.* gopher://israelinfo.gov.il:70/00/govmin/envir/envlaw/frwater/941032.evl.

Kay, P. and B. Mitchell (2000). Water Security for the Jordan River States: Performance Criteria and Uncertainty. In: *Water in the Middle East*, Hussein A. and A. Wolf, eds. Austin: University of Texas Press.

Kolars, J. F. and W. A. Mitchell (1991). *The Euphrates River and the Southeast Anatolia Development Project.* Southern Illinois University Press.

Lewis, B. (1991). "After the Gulf Crisis." In *Studies on Turkish Arab Relations* no. 6, pp. 25–81.

Lithwick, H. (2000). Evaluating Water Balances in Israel. In: *Water Balances in the Eastern Mediterranean*. Brooks, D. and O. Mehmet, eds. Canada: International Development Research Centre.

Lovatt, D. (2002). *Turkish Foreign Policy Towards Iraq and Syria since 1990; Syria-Iraq rapprochement.* http://www.ir.metu.edu.tr/conference/papers/lovatt.pdf.

Lowi, M. (1993). *Water and Power: The Politics of a Scarce Resource in the Jordan River Basin.* Cambridge: Cambridge University Press, p. 108.

Medzini, A. (2001). *The Euphrates: A Shared River,* SOAS Water Research Group of the School of Oriental and African Studies, University of London.

MEI (1990). "Turkey, Syria and Iraq." *MEI,* 16 February, p. 12.

MEI (1993). "Syria: Politics, the Economy and the Succession," *MEI,* 16 April, pp. 16–17.

Middle East (2002). "Oil Exports to Syria Help Keep Baghdad Afloat." *Middle East* (June 2002), Issue 324, p.23.

Naff, T. and Matson, R. C. (eds.) (1984). *Water in the Middle East: Conflict or Cooperation?* Boulder: Westview Press.

Nasrallah, F. (1990). "Middle Eastern Waters: The Hydraulic Imperative." *Middle East International,* 27 April, no. 374, pp. 16–17.

Portnov, B. (1988). "The Effect of Housing Construction on Population Migrations in Israel." *Journal of Ethnic and Migration Studies,* 24:(3), p. 541.

Qubain, F. I. (1960). *The Reconstruction of Iraq: 1950–1957.* London: Atlantic Books, Stevens & Sons.

Rogers, P. (1991). *International River Basins: Pervasive Unidirectional Externalities.* Presented at a conference on The Economics of Transnational Commons, University di Siena, Italy (April) pp. 25–27.

Schiffler, M. (1998). *The Economics of Groundwater Management in Arid Countries: Theory, International Experience and a Case Study of Jordan.* London: Frank Cass Publishers.

Shannag, E. and Y. Al-Adwan (2000). Evaluating Water Balances in Jordan. In: *Water Balances in the Eastern Mediterranean*, Brooks, D. and O. Mehmet, eds. Canada: International Development Research Centre.

Tekeli, S. (1990). "Turkey Seeks Reconciliation for the Water Issue Induced by the Southeastern Anatolia Project (GAP)," *Water International*, 15, pp. 206–216.

Tuma, E. H. (1986). *Economic and Political Change in the Middle East*. Palo Alto, CA: Pacific Books.

Turkish Embassy, Washington, DC (1996). *Syrian-Turkish Water Issues, Background Note*, June 1996. http://www.turkey.org/releases/062096a.

U.S. Department of State (1995). *Background Notes, Jordan*, Accessed Apr 15, 2000. http://www.state.gov/www/current/middle_east/isjordan.html.

U.S. Geological Survey (1998). *Overview of Middle East Water Resources: Water Resources of Palestinian, Jordanian, and Israeli Interest*. Compiled works of the Jordanian Ministry of Water and Irrigation, the Palestinian Water Authority, and the Israeli Hydrological Service. Washington DC.

Wolf, A. (1998). "Conflict and Cooperation Along International Waterways." *Water Policy*. 1: (2), pp. 251–265.

———. (1999). "Criteria for Equitable Allocations: The Heart of International Water Conflict." *Natural Resources Forum*. 23: 1, February 1999. pp. 3–30.

———. (1995). *Hydropolitics along the Jordan River: Scarce Water and its Impact on the Arab-Israeli Conflict*. Tokyo: United Nations University Press.

———. (2000). "Hydrostrategic" Territory in the Jordan Basin: Water, War, and Arab-Israeli Peace Negotiations. In: *Water in the Middle East*, Hussein, A. and A. Wolf, eds. Austin: University of Texas Press.

Chapter 5

The Aral Sea Basin: Legal and Institutional Aspects of Governance

Laurence Boisson de Chazournes

1. THE ARAL SEA CRISIS

1.1. The Biophysical Dimension

Once the world's fourth-largest inland body of water, the Aral Sea, has shriveled to occupy half its former area and to a third of its volume. The Aral Sea fell victim to a drive led by Soviet planners, who decided in the 1960s to make their country self-sufficient in cotton and to provide employment for a rapidly growing population. Irrigation water was thus diverted from the Amu Darya and from the Syr Darya rivers flowing into the inland sea. This diversion of water accentuated the shriveling of the sea by natural processes. As a result, besides the disappearance of the water itself, disaster conditions also spread to large portions of the ter ritories upstream of the Aral Sea: productive wetlands in the deltas have dried up; salt dust from the bed of the drying sea and chemicals from fertilizers used in cotton fields have endangered the health of millions of people living in the region; in addition, salinity in the rivers has risen to very high levels, entailing serious threats to soil productivity.

President Mikhail Gorbachev's ascension to power in 1985 brought to light the dimensions of the crisis.[1] The Aral Sea was nearly biologically dead and rapidly diminishing in size, while formerly productive land was dead or dying. On attaining independence in 1991, the water management system and associated investments and transfers provided by the central planning system in Moscow were gone, and the potential for conflict was very high. The five new Central Asian republics (Kazakhstan, the Kyrgyz Republic, Tadjikistan, Turkmenistan, and Uzbekistan) thus recognized the urgency to react to this issue.

147

1.2. The Historical and Political Context

With the demise of the former Soviet Union in December 1991, five new countries emerged in Central Asia. Together with Afghanistan, they became riparians to the Amu Darya and Syr Darya rivers, which are two important international watercourses. This new political situation meant that paradigm shifts had to take place: the recently established States had to manage two international watercourses in accordance with principles and rules of international law in the context of their new mutual relationships, relationships that were no more of a domestic nature as they had been during the Soviet period. Moreover, the central planning and management system for energy, food, and other economic activities, put in place during the Soviet era was suddenly withdrawn and had to be replaced by a regional cooperative system between the riparian States. In addition, the five countries, regarded by the central Soviet planners before 1991 as a single agricultural region for economic purposes, have since developed different views with respect to water uses. Hence, the upstream States claim water for hydropower uses, while the downstream States mainly rely on water for agricultural uses.

The collapse of the former Soviet Union also brought to the fore an ecological catastrophe of great magnitude, the disappearance of the Aral Sea. The prevailing ecological system has its roots in mismanagement and diversion of rivers flowing into the Aral Sea. The present actors have inherited the pillars of the previous allocation system and have to work on these bases when renegotiating their agreements and putting in place joint institutions. The actions adopted so far regarding the Aral Sea Basin are set against this background, namely that a sustainable water management system based on an adequate framework of benefit to all riparians should be established. Several legal and institutional steps have been taken in this direction, although there is a need for further action to elaborate viable instruments based on a common understanding of all five riparians. These accomplishments constitute confidence-building measures that are necessary in order to reach agreement over a wide array of issues, some more contentious than others.

The Aral Sea crisis was to be dealt with at a basin level, taking into consideration the particularities of the Aral Sea Basin.[2] The regional dimension was thus defined by the drainage future of the area. The Aral Sea is a closed drainage area, the characteristic feature of which is a marked variety of relief forms. Plains cover its western and central parts; the southern part is occupied by large mountain ranges. The rivers of the basin are fed predominantly by snowmelt, glacier-melt, or both. Rivers flow from the mountains onto the plains and are mainly exhausted and disappear in the deserts' sands, with the exception of the two largest, the Amu Darya and the Syr Darya, which cross the deserts and flow into the Aral Sea. The Amu Darya rises from Tadjikistan and Afghanistan and flows through Uzbekistan and Turkmenistan to the Aral Sea. The Syr Darya rises in the Kyrgyz Republic and flows through Tadjikistan, Uzbekistan, and Kazakhstan into the Aral Sea.[3] Except for Kazakhstan, about 90 percent of the territories of the four other countries are within the basins of the two main rivers.

The Aral Sea crisis has to be seen in a broad context, which includes the management of water and agriculture in the entire Aral Sea Basin, the root cause of the problems. The interactions between physical, environmental, economic, political, and social factors must be taken into account, as they all play a role in the crisis.[4] It is interesting to note that while these newly independent States made arrangements for water distribution, they did not make such agreements regarding other resources such as oil and gas. Nonetheless, differences of opinion have emerged with respect to water, and notably as regards its scarcity. It should also be noted that the focus of attention, which in the past was the depletion of the Aral Sea, has moved toward better management of the international watercourses. Kazakhstan seems to remain the only country that shows some interest in the preservation of what remains of the Aral Sea at least of its northern part, known as the "Little Aral Sea."[5]

In particular, Kazakhstan has worked with the World Bank for several years in order to develop and implement a lasting strategy to improve water management and to rehabilitate the Northern Aral Sea (NAS). The latest project deriving from the cooperation with the World Bank was approved in 2001 and is expected to cost about $86 million of which 75 percent will be financed by the World Bank (IBRD). In its first phase, the project aims at rehabilitating the NAS by building a dike across the channel connecting the NAS to the larger Aral Sea. It should also improve the hydraulic structures on the Syr Darya, as well as rehabilitate a dam (the Chardarah dam), restore aquatic resources, and develop fisheries.[6]

2. ANALYSIS OF RELEVANT INSTITUTIONS AND ACTORS

2.1. Identification of the Relevant Institutions and Actors

2.1.1. A Plan for Action: The Aral Sea Basin Program

At the request of the five Central Asian republics, a number of international and organizations such as UNDP, UNEP, the World Bank, and the European Union (hereafter the donors) provided support to enable these countries to elaborate ideas for long-term solutions. This culminated in the adoption by the five Central Asian republics of a comprehensive Aral Sea Basin Program (ASBP) in January 1994.[7] The ASBP attempts to deal in a comprehensive fashion with the full range of problems inherent in the Aral Sea crisis. It is a transboundary and multi-sectoral agreement, including measures devised to develop sustainable water and related land resources management strategies, improve the information system needed for all planning and management activities, mitigate the impacts of environmental degradation, ameliorate the conditions in the upper watersheds and the area adjacent to the sea, and strengthen the implementation capabilities of the competent regional institutions.

The ASBP has four long-term objectives: (i) to stabilize the environment of the Aral Sea Basin; (ii) to rehabilitate the Disaster Zone around the sea; (iii) to improve the management of the international waters of the basin; and (iv) to build the capacity of regional institutions to plan and manage these programs. It is also intended to assist the riparian States in cooperating and adopting sustainable regional policies to address the crisis, as well as to provide a framework for establishing national macroeconomic and sectoral policies with a view to achieving sustainable development of land, water, and other natural resources.

It was recognized at the outset that achieving these objectives—which in fact were supposed to reverse forty years of water resource mismanagement and environmental destruction—would be an enormous undertaking and would have to be approached in gradual phases. The first phase of the program, placing emphasis on assistance to the people of the Disaster Zone, and building up the knowledge base and the institutions required to deal with longer-term issues, is to be completed in three to four years. In the next stage of the program, attention should be focused on a few strategic regional water management problems, and on intensifying complementary efforts at the national level, both to meet the needs of the people of the Disaster Zone and to promote sustainable resource use in the basin. Stabilizing the environment, rehabilitating the disaster zone around the Aral Sea, improving the management of international waters, and building the capacity of the regional institutions were identified as core elements of the core regional program. The last stage, which should be completed by 2025, should expand and continue the program already undertaken.

The heads of States have met at least once a year in the past eight years to develop, approve, and express continued support for the program. Moreover, the basin State governments have acted to realize watershed-wide gains where it was clear how to do so. An Immediate Impact Project was added to the ASBP in 1995, in order to meet the needs of people in the Disaster Zone as quickly as possible.

Among the objectives to be achieved through the ASBP, was the improvement of the management of the waters of the basin as well as the strengthening of the capacity of institutions in charge of this management. The elaboration of an appropriate institutional and regulatory framework was a means to this end.

2.1.2. The First Legal and Institutional Steps

Committed to avoiding conflict over water issues after their independence, the five Central Asian States immediately put in place interim arrangements and institutions for water sharing. They were aware of the need to elaborate an appropriate institutional and regulatory framework to deal with water scarcity issues at an intergovernmental level. All five countries were going through dramatic economic changes and their claims on water uses were to be seen within the framework of these changes, each having different needs in the use of water, be they oriented toward irrigation needs or hydropower developments.

2.1.2.1. The 1992 Agreement on Cooperation in the Management, Utilization, and Protection of Water Resources in Interstate Sources

During the former Soviet era, inter-Republican water resources were administered centrally under the aegis of the Ministry of Water Management. Schemes for water uses were developed for the Syr Darya and the Amu Darya River basins based on annual water withdrawal limits. They were devised on the basis of crop requirements and little attention was paid to water quality. Due to the seasonal variations, the republics would enter into a series of bilateral and trilateral agreements to correct water allocations made on the schemes.

At the time of their independence, all five countries formulated claims to an equitable share of the waters, acknowledging at the same time that this could only be achieved through international negotiations. Ultimately, it was agreed that the status quo would be maintained and would be an equitable formula for the time being. This was achieved through the five Central Asian countries jointly declaring on September 12, 1991 that mutual water resources management would be a basis for equity and joint benefits. They subsequently concluded the Agreement on Cooperation in the Management, Utilization and Protection of Water Resources in Interstate Sources on February 18, 1992. This constituted the first step toward the establishment of a cooperative scheme: the riparian States thereby formally acknowledged their commitment to implement the cooperative management of waters in the Aral Sea Basin.

Under this agreement, the five States realized that they have common interests in the use and protection of shared water resources and equal rights and responsibilities in this respect. The water resources of the region were defined as "common and integral." The republics codified past practices in promising to provide strict observance of the order relating to water allocation established and agreed upon during the Soviet period. They also committed themselves to refrain from conducting activities that would result in a deviation from the agreed water shares, from causing water pollution, or from allowing any deviation likely to affect detrimentally the interests of the five States. They also agreed to carry out joint activities for the solution of the Aral Sea crisis and to determine yearly sanitary releases based on water availability for the Aral Sea.

2.1.2.2. The Establishment of Regional Institutions

As a first institutional step, the five States established the Interstate Commission for Water Coordination (ICWC) in 1993. Its members, who meet several times per year, are the heads of the main water management organizations in each country (or their designate). The commission is the institutional framework for water management, including water allocation issues and the approval of schedules for the operation of reservoirs. Its decisions are adopted by unanimous vote and are binding on all water users. It should also be mentioned that two river basin agencies, the water management authorities (BVOs), were established by the Ministry of Water Resources of the USSR in 1986, for each of the two rivers. They are vested with executive functions with respect to the operation of hydraulic structures and installations on the rivers and are in charge of the management and monitoring of the allocation made by the ICWC to member States. They are among the symbols of continuity with the Soviet era, carrying with them the emphasis that was put on quantity aspects as well as on the economic water uses at stake. Both the ICWC and the BVOs are responsible for ensuring compliance with water withdrawal limits and for guaranteeing the annual volume of water to be supplied to the Aral Sea and to its deltas. The ICWC and BVOs are directly responsible to the States for the execution of their functions. The countries have obligations regarding financial and technical support for both BVOs.

To further the reinforcement of cooperation among the five States, four other intergovernmental institutions were created between 1993 and 1995. These are (i) the Interstate Council on the Aral Sea Basin (ICAS) intended to set policy, provide intersectoral coordination, and review the projects and activities conducted in the basin; (ii) the Executive Committee of ICAS (EC-ICAS), a secretariat responsible for implementing the Aral Sea Basin Program; (iii) the International Fund for the

Aral Sea (IFAS), whose purpose was to collect contributions for ICAS from the five States and donors; and (iv) the Sustainable Development Commission (SDC), established to ensure that economic, social, and environmental factors are given equal weight in planning decisions. The SDC was intended to function within the framework of ICAS and to complement the input of the Water Ministries by defining proposals for ICAS addressing the ecological protection and socioeconomic development of the basin.[8]

These new institutions helped strengthen the willingness, and more importantly, set the framework, for the five countries to decide jointly on water management issues, an exercise in which the donor community was involved through the provision of technical assistance.[9] However, to rationalize the allocation of responsibilities and streamline the decision-making process in the Aral Sea Basin,[10] a modification of this institutional structure was deemed necessary. Being concerned with the establishment of an effective donor grants management process, such changes were held as particularly desirable to the donors.

Following a review of the ASBP made by the World Bank and by other donors, a new fund was created, the International Fund for the Aral Sea (IFAS). It is a successor to the former ICAS as well as to the former structure of the IFAS. Established in 1997, it has a board composed of the deputy prime ministers of the five States concerned with agriculture, water, and the environment. The board meets regularly to reconcile the views of member States and decides on the policies, programs, and institutional proposals recommended by the Executive Committee (EC) (renamed the Executive Committee of the IFAS). Moreover, the IFAS collects contributions and finances program activities.

2.2. Relationships and Interconnections between the Different Institutions: Interstate Water Resources Management Structure in the Aral Sea Basin

The relationship between this new IFAS and the International Commission for Water Coordination (ICWC) was further clarified through an agreement concluded between the heads of States on April 9, 1999. As a result, responsibilities have been distributed in the following manner:

Board of the International Fund for the Aral Sea (Board of the IFAS)—comprised of the deputy prime ministers of the five States. It is the highest political level for decision making and final approval of activities before (if needed) approval by the heads of States;

Executive Committee of the IFAS—a permanent body, comprised of two members from each State, carrying out all activities for the implementation of decisions made by the board of the IFAS via the national branches of the IFAS. Also, on behalf of the board, the EC-IFAS coordinates the

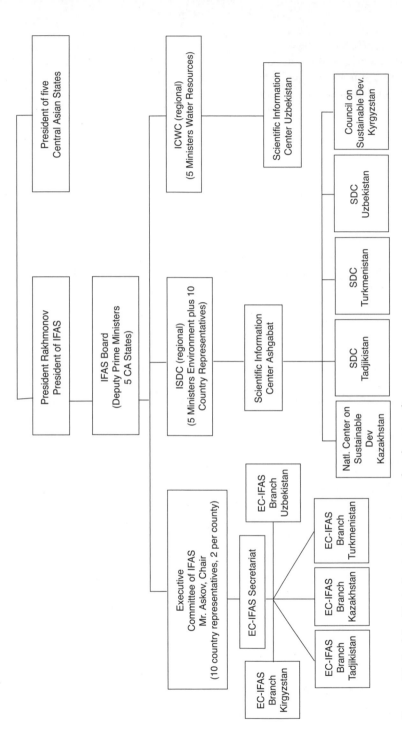

FIGURE 12. Aral Sea Basin Institutions (regional and national)

activities of the agencies involved in the implementation of projects (international and donors);

Interstate Commission for Water Coordination (ICWC)—the highest level of transboundary water resources management, water allocation, water monitoring, water use, and preliminary proposals assessment for principal improvement and change of organizational, technical, financial, environment policies, and decisions related to water on the interstate level. BVOs and the Scientific Information Center (SIC ICWC) are executing bodies of this commission, which is not a permanent body.

Interstate Sustainable Development Commission (SDC)—the regional SDC consists of three representatives from each of the five States, one from each of the following three bodies: the ministers of Nature Protection/Ministry of Natural Resources; representatives of the national SDC; and representatives of the Ministries of Economy/Ministry of Macro-Economy. The regional SDC has a six-year mandate, but intends to submit an updated version of its mandate to the board of the IFAS for its consideration.

3. MANAGEMENT OF THE ARAL SEA BASIN: CURRENT ISSUES AND ISSUES AHEAD

The Central Asian countries, together with the World Bank, the Global Environment Facility (GEF), and other donors (the Netherlands, the European Union/TACIS,[11] the Swedish International Development Agency [SIDA], and the Swiss government [SECO]), have developed a project[12] aiming at substantially contributing to the stabilization of the environment and to the improved management of the waters. This Water and Environment Management Project (WEMP) achieves its aims by promoting coherent national and regional water and salt management policies. Attention is also given to education of the general public as to water conservation, including the possible need to accept difficult decisions in the shorter-term for longer-term gains. The WEMP also focuses on improving the management of reservoirs and dams as well as the monitoring of transboundary water flows—the latter both in terms of quantity and quality. Finally, it plans to rehabilitate a pilot wetland in one of the Aral Sea deltas.

To complement the WEMP, the U.S. Agency for International Development (USAID) has initiated the Natural Resource Management Project (NRMP), which provides technical assistance in improving management of critical natural resources in the five countries of central Asia. With respect to its water related activities, the NRMP is designed to increase water management capabilities in the region through training programs, upgrading data management, improving the understanding of the implications of natural resource policies and regulations, strengthening skills for design and implementation of demonstration projects, and also increasing public awareness of environmental issues.[13]

Additionally, the United Nations Educational, Scientific, and Cultural Organization (UNESCO) is becoming increasingly active in the field of water conflicts prevention. Hence, UNESCO (in collaboration with Green Cross International ([GCI]) is at the head of the PC->CP initiative (from Potential Conflict to Co-operation Potential), whose purpose is notably to define and survey potential conflicts arising from water resource management and to provide concerned States with water-related negotiations and cooperation-building techniques. The Aral Sea Basin is expected to be one of the first cases to be examined by UNESCO under that project.

3.1. The Need for Further Development of the Water Management Framework

Among the issues to be dealt with is the need to clarify the legal framework for water management. Although the five States have committed themselves to the water-sharing order agreed under the Soviet era, there is still scope for uncertainties and diverging practices that may prove to be a source of conflict: criteria for water sharing are not expressly stated; water resources conservation and planning are not clearly envisaged; the problem of reservoirs and the economic and social needs at stake in the region are potential irritants; and water is still used fairly inefficiently. It has also become increasingly apparent that the quantifiable minimal flow of water to the Aral Sea would have to be formulated so as to preserve what remains from this sea. Finally, in the event of a conflict, there is no adequate dispute settlement mechanism.

This situation was notably identified by the World Bank's WEMP and by the USAID's NRMP, as well as by the European Union (TACIS Program) as a possible area where the donor community could play a role in providing technical and financial assistance.

In particular, the EU's TACIS Program includes a section on Water Resources Management and Agricultural Production (WARMAP) in the Central Asian republics, whose general objectives are (i) to provide the administrative and technical framework within which policies, strategies, and development programs for utilization, allocation, and management of the water resources of the Aral Sea Basin can be developed; and (ii) to assist at the regional level with the establishment of the institutional structure required to prepare and implement the policies and strategies on water allocation and management. Among the specific objectives is the provision of a legal basis for international and national water resources utilization, giving due weight to the environmental needs of the Aral Sea Basin. It was agreed by the five States that the TACIS Program[14] would support the drafting process of water-sharing agreements. The program was launched in 1995, based on the organization of training activities, the setting up of working groups, and the advice provided by experts on international water law issues.

To date, three draft agreements have been produced, covering the functions, powers, and responsibilities of the Interstate Commission for Water Coordination and its organs, as well as the general principles applicable to the Aral Sea Basin in its entirety. They also provide a mechanism for joint interstate strategic planning of the management, development, and protection of transboundary water resources.[15]

Although these draft agreements lack precision, they nevertheless provide for quantifiable minimum releases of water to the Aral Sea and to its deltas. They also set a framework for information sharing on the planning of activities between States and prescribe respect for principles of international law as reflected in the UN Convention on the Law of Non-Navigational Uses of International Watercourses,[16] such as the principle of equitable and rational use, the "no-harm rule," and the obligation to cooperate. Yet the draft agreements neither further specify how these principles should be applied, nor provide how they should interact with each other.

According to them, the ICWC is in charge of allocating annual limits of water to be used by the parties. They also intend to supplement these provisions with two attachments, the first one dealing with the Amu Darya Basin and the second with the Syr Darya Basin. They will contain water allocation criteria and operational regulations for each of the river basins.

Another step to promote a more efficient water management framework was to organize, at the domestic level, working groups composed of specialists in water, energy, and agricultural issues, and to establish regional working groups to work on drafts before their submission for adoption to the governments. The EU TACIS Program continues to support these processes, but nevertheless, the conclusion of the agreements will depend greatly on the political determination of the States.[17] At present, the riparian States do not see this outcome as a priority. Preference seems to be given to developing national and regional strategies for water use in the Aral Sea Basin, rather than to commit to stringent water allocation quotas. This is seen as a better way to ensure that all of the States get on board and that their levels of water consumption are reduced.

Although it may not prove to be very fruitful in the short-term as a means of arriving at the treaties' signature stage, a virtue of the legislative process may nonetheless be that the creation of working groups forms part of an educational and information-sharing process. Indeed, it helps to consolidate the understanding among the various stakeholders, an element as important in the negotiation and adoption of water agreements as in the verification of their effective implementation.

Another matter of concern relates to the quality of the transboundary waters. It was decided that this issue would be negotiated in a separate agreement. However, this area is in a crucial need for initial action, since there is at present no water quality management scheme. Attention has to be paid to key pollutants such as salts, as well as to monitoring and control. It is also important that water quantity and quality problems are dealt with together, as they are physically linked and intertwined at a managerial level.

The possibility of Afghanistan becoming a party to any relevant agreement should also be taken into consideration. Afghanistan, an upstream riparian of the Amu Darya River, may decide to develop water resources for its own use.[18] The involvement of Afghanistan is therefore important for ensuring effective long-term management of the waters of the Aral Sea Basin. From a legal standpoint, such participation should be seen as pursuing the process embedded in the treaties concluded between Afghanistan and the former USSR, to which the five Central Asian republics are successors.[19]

Under the technical assistance deriving from the second phase of the WARMAP Project of the EU-TACIS Program, the BVOs have created their own database, which is compatible with the regional Information System already put in place. The work concerning further development of the basin-wide Information System should be continued and financially supported by the donor community. As part of this work, modeling of the operational management should take place.

Recently, the ICWC has been in charge of developing a water management framework through its collaboration with the International Water Management Institute (IWMI) (and its partner in this project, the Swiss Development Council). Indeed, the IWMI established an office in Tashkent (Uzbekistan) in 2001 and has been notably focusing its research on building strong water management institutions, essentially by endeavoring to develop a framework to transfer water management responsibilities from agencies managing water along administrative boundaries to an institution managing water along hydrological boundaries.[20] Simultaneously, the IWMI concentrates its effort on improving water productivity by identifying the best practices in order to reduce the amount of water used by farmers.

3.2. The Need to Improve the Management of the Water Storage System

A complex system of water storage was built during the Soviet period in the Syr Darya and Amu Darya river basins, with the initial primary purpose of accumulating water in winter for its subsequent use in the summer, mainly for irrigation and electric power production.[21] Downstream States benefited by gaining a supply of hydropower, while upstream States received energy supply in the winter. Yet problems arose with respect to financing and responsibility for the operation of the infrastructure, problems that were partially solved during independence when it was decided that the infrastructure would be owned by the State in which it is located but that management activities would be shared by the States and by relevant river basin authorities. This nonetheless leaves uncertainties with potential for disagreements in the longer-term, in particular as local ownership is not conducive to the coordination of regional water use requirements.[22] Due to seasonal variations, the States have established a practice of concluding bilateral and multilateral agreements to correct the water allocations.[23]

A special case is the Toktogul hydro-power station and reservoir located in the Kyrgyz Republic. The storage facility (19 km^3) controls downstream water use in Uzbekistan and in Kazakhstan. This situation called for a more formalized agreement, with all interested States agreeing on a formula to be codified so that disputes can be avoided.

Following independence, the Kyrgyz Republic found itself with an abundance of water and hydroworks and with a severe shortage in energy supply during the winter. The downstream States, Uzbekistan and Kazakhstan found themselves with an agriculture-based economy with water demands during the summer months but with an upstream neighbor that needed to release the waters during the winter period in order to supply its energy needs. In order to resolve this matter, Uzbekistan, Kazakhstan, and the Kyrgyz Republic entered into a series of annual agreements that tend to establish trade-offs between energy and water: the Kyrgyz Republic agreed to reduce water releases during the winter months in return for Uzbekistan and Kazakhstan supplying it with electricity and fossil fuel during the winter (i.e., water storage for summer irrigation/energy production and delivery of coal and gas in the winter).

There was a need to settle the operation of the reservoir in a more predictable way so that all of the countries could meet the water management objectives. This would place the Toktogul Reservoir on a secure footing, eliminating an important potential source of conflict over water.

Yet another international institution, the Interstate Council for Kazakhstan, the Kyrgyz Republic, and Uzbekistan (ICKKU)—now known as the Central Asian Cooperation Organization (CACO)—was established. With the assistance provided by the USAID, it has negotiated a

Multi-Year Interstate Agreement on Management on the Naryn-Syr Darya Cascade among the four riparian States (Uzbekistan, the Kyrgyz Republic, Kazakhstan, and Tadjikistan). The agreement would encompass regulation of the timing of water storage releases from the Toktogul reservoir of the Naryn-Syr Darya Cascade through compensatory schemes based, inter alia, on seasonal water storage and delivery of coal and gas. It would also take into account the issue of the valuation of the price of water. The process was launched at a meeting of high-level government officials of the States involved in December 1996. In March 1998, Uzbekistan, the Kyrgyz Republic, and Kazakhstan signed a framework agreement acknowledging the principle of financial compensation. It was also agreed that Tadjikistan would be invited to join the agreement, which it did in 1999. Following the signing of this agreement, USAID management activities led to the use of a planning tool (developed by the latter) enabling the parties to make decisions on the allocation and distribution of water in the region, as well as to the creation of another water management facility on one of the Syr Darya's major tributaries, the Chirchik River.

Several issues thus remain at stake. For instance, the USAID is expected to continue providing technical assistance to further develop the technical basis of the agreement. However, some changes in the ongoing activities are anticipated after an internal assessment of the new measures' outcome. One issue concerns the choice of the regional institution taking charge of such processes, as several institutions (mainly IFAS and CACO) are involved. The answer to such questions rests on the trust placed by each country in the processes and on the political resolve and determination they will demonstrate to set up an adequate framework. The donor community plays an important role in the choices to be made and its influence should be supportive of solutions satisfying the interests of all of the partners. The proliferation of competent institutions is not desirable in a context where there is a need for a more integrated approach, both in a technical and political sense.

3.3. The Need for Integrating Water, Energy, and Trade Concerns

Between 1985 and 2002, the volume of cross-border power transactions through the integrated water and power systems of the five Central Asian countries dropped from $15 billion a year to $1 billion. An insular approach to energy self-sufficiency and a dramatic reduction in maintenance spending has created a perverse result. In spite of growing electricity shortages in the region, water that exceeds the storage capacity of mountain reservoirs is often released without producing a single kilowatt-hour of electricity. In this context, USAID experts provided technical advice and negotiation support to the Asian Development Bank (ADB) in the successful negotiation and adoption in November 2002 of a Power Trade Relations Agreement between the governments of Uzbekistan and Tajikistan.

The agreement establishes a framework for bilateral power-trade relations to reduce the combined electricity costs of the two countries by $20 million a year in the near term. The agreement also specifies the policy and institutional conditions needed for an integrated water and energy system in the whole of Central Asia. Such a regional system would enable the dispatch of energy and water resources on the basis of economic criteria and has a potential windfall of about $1 billion a year. The agreement will reduce electricity costs, improve the use of the region's water resources, and help to reduce potential conflicts over water and energy resources. It is expected that the other Central Asian countries will join the agreement.

Successful completion of the agreement was a precondition of a $175 million loan package to be provided by the ADB and by the European Bank for Reconstruction and Development (EBRD) to upgrade the electric transmission networks of the two countries. The renovation of the physical network and the reorientation of the regional water and energy systems along market principles promises to benefit the entire region with savings that will flow to all of the consumers, including significant benefits to low-income customers. Currently detailed work plans, cost estimates, and implementation arrangements are being prepared in order to devise various provisions in the agreement. These plans were discussed at the first meeting of a regional working group in January 2003 with support from USAID's Transboundary Water and Energy Project.[24]

3.4. How the Situation Is Likely to Evolve

3.4.1. Technical and Scientific Activities As Crucial Components for Developing and Implementing Legal Instruments and Strategies

Over the medium-term, water quality is emerging as an issue of great significance, especially with respect to salinity. Indeed, it is now considered that salt management will be the major challenge to basin water resources management for the next decade. The rivers transport 140 million tons of salt per year, induced partly by natural runoff, but mostly by extensive irrigation development. As a result of insufficient drainage in some areas and excessively deep drains in others, salt mobilization from the soil profile is much greater than the internationally accepted practice. If no action is taken, the problem of severe salinity will most likely spread from the lower watersheds to the highly productive middle watershed areas.

The salinity problem is complex: there is neither a single source, nor a single solution. In this context, WEMP, funded by the World Bank, by the GEF, and by a few other donors, is intended to provide a basis for technical assessment, allowing the collection and analysis of relevant data with a view to the development of a salinity strategy. Related investments of $65 million dedicated to improve water management and water conservation—for example, irrigation improvements in Kazakhstan, Uzbekistan, and the Kyrgyz Republic—are associated with this endeavor.

The proposed strategy is expected to provide the basis for the negotiation of a legal agreement in which the Central Asian republics would deal with the salinity management issue. The proposed project will help determine salinity standards and the locations along the two rivers where salinity levels might be controlled. It will also help the various States assess the costs of dealing with different salinity levels that may potentially occur. The latter is important, as the downstream States will be the main beneficiaries of a salinity management scheme while the upstream States will have to take stringent measures for addressing the problem. Ultimately, the States could also decide to revise their shares of water accordingly.

The WEMP and related activities reveal one of the roles that financial and technical assistance can play with respect to institutions and regulations.[25] Such assistance can support the conduct of scientific and technical activities that are of significant importance to the design of a legal regime, for they help in identifying and remedying problems. Financial and technical assistance may open paths leading to the negotiation of international agreements; it can also promote the edification of mechanisms that monitor the regime put in place and allow for its adaptation to new needs. In this respect, financial and technical assistance are important components in a process aimed at negotiating viable instruments.

3.4.2. Public Participation and Other Institutional Changes

Government plays a predominant role in water resources development and management in Central Asia, and it has been asserted that political leaders have shown to be wary of public participation and direct influ-

ence; that the bureaucracies tend to oppose sharing management decisions or being subject to oversight and the public at large often lacks
information; and, that implementing agencies were never accountable to
the public for their activities.[26] In most cases the public was only periph-
erally involved in the guise of special interests and pressure groups. The
growing importance of public opinion and NGO activities has been
recognized only since the midnineties.

One of the centerpieces of the WEMP project is a Public Awareness Component, whose objective is to heighten the public's awareness of water conservation issues. The project has two phases. The
first is to draw to the attention of decision-makers the importance
and indeed financial benefits, of taking into account public opinion
about water resources management and development and the need to
establish a strategy to improve water management. In the second
stage, governmental institutions and implementing agencies should
provide access to information about their policy and strategy for the
public at large. The public should become an advocate for the proposed strategy's implementation.

The water sector has always played a vital role in the region.
To increase human potential, especially at the level of decision making in the water sector, the public awareness campaign and nongovernmental organization (NGO) involvement have a crucial role.[27] At
the regional level, public participation is understood as a broad
spectrum of information activities about regional organizations' activity through EC-IFAS in close collaboration with water-related and
environmental NGOs. The first meeting of NGOs and IFAS organizations took place in May 1999 and a memorandum about mutual
activity was agreed upon.

4. CONCLUDING REMARKS

At the end of the 1980s, the Aral Sea crisis attracted international
attention. The five new States emerging from the ruins of the Soviet
Union were committed to acting together to face the catastrophe. While
codifying past practice agreed under the Soviet period in terms of water
allocation, they recognized the need to strengthen the existing institutional and regulatory framework and to adapt it to their new demands. They have established a relatively comprehensive framework,
which makes it fairly unique among the river basin organizations.
They now need to show political commitment to implement effectively

this endeavor, a commitment which, for the time being, is rather eva-
nescent to say the least.

The donor community has provided assistance in achieving this
aim. The European Union, UNDP, the World Bank, and other donors
have given substantial support for capacity building at the regional
level. However, for financial and technical activities to reach their objectives, a key aspect is coordination among donors. This is particularly
true for the design and implementation of an adequate international
institutional and regulatory framework. An additional issue is for an
international organization to take the lead for coordinating donors'
activities. Once at the forefront, the World Bank seems to be disengag-
ing itself from this role. This raises problems for creating incentives at
the regional and local levels for implementing sound water management
policies and activities.

At the institutional level, there is a need to clarify further the
relationship between the various regional organs (IFAS and ICWC) as
well as with CACO. This would strengthen both the decision-making
and implementation processes. It is important that all parties perceive
that their interests are taken into consideration. Enhanced consultation
and joint action are based on these premises. The institutional setup
also needs to be refined so that it reflects the need for integrated management, in particular for managing together water quantity and quality
problems. Public participation at all levels of water management should
also be encouraged, although it should be recognized that for the region, it is a new issue to be dealt with and that today, such a process
remains in a nascent phase.

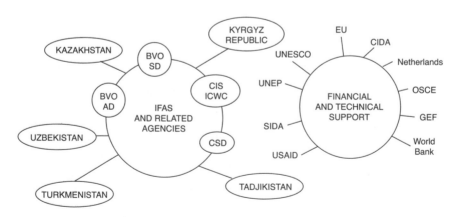

FIGURE 13. Regional and International Organizations in the Aral Sea Basin

NOTES

Director and Professor, Department of Public International Law and International Organization, Faculty of Law, University of Geneva, Switzerland. The author would like to thank Anatoly Krutov, Operations Officer, World Bank, for his comments on an earlier version of this book. However, the ideas expressed in this chapter are those of the author.

1. See Philip Micklin, "International and Regional Responses to the Aral Crisis: An Overview of Efforts and Accomplishments," 34 *Post-Soviet Geography and Economics*, 399, pp. 403–404.

2. European Commission (TACIS), *WARMAP Project: Formulation & Analysis of Regional Strategies on Land and Water Resources* (1997), p. 7.

3. The Aral Sea's drainage basin covers about 1.8 million km^2 in seven States (five republics of the former USSR, Afghanistan, and Iran). Only about 0.5 million km^2 of this area actively produces or consumes water that could enter the Aral Sea. Iran's contribution to the flows in the basin is entirely in streams ending in the Kara-Kum Desert that cannot actually reach the Aral Sea. Ibid.

4. See Micklin, *supra,* 403. On the environmental dimensions of the disaster, see M. H. Glantz, "Creeping Environmental Problems in the Aral Sea Basin," *Central Eurasian Water Crisis: Caspian, Aral and Dead Seas*, Kobori, I. & M. H. Glantz (eds.), pp. 25–52 (Tokyo: United Nations University Press, 1998).

5. "The Aral Sea: Saving the Last Drop," *The Economist,* July 1, 2000 at p. 64

6. See the World Bank's website, at http://www.worldbank.org/projects (last visited September 2002). Simultaneously and also through the World Banks' assistance, Kazakhstan is undertaking an irrigation and drainage improvement project, whose main objectives are to promote sustainable agricultural production and practices, notably through the rehabilitation of its irrigation and drainage systems.

7. In June 1994, a donors meeting was held in Paris, and support for some elements of the first phase of the program (ASP-1) was pledged by various multi- and bilateral donors. The program has enjoyed solid donor support, backed by $32 million in grant financing pledged in Paris in 1994 for the preparation stage and now committed. For the current amounts of donor contributions, see Table 10.

8. See the Regional Environmental Action Plan for Central Asia, adopted in September 2001 and prepared with the support of UNEP, UNDP, and the Asian Development Bank.

9. On institutional aspects, and more generally on the bank's role in the Aral Sea crisis, see Kirmany, S. & Moigne, G. L. *Fostering Riparian Cooperation in International River Basins—The World Bank at its Best in Development Diplomacy 10–15* (World Bank Technical Paper no. 335, 1997).

10. On the need to improve the institutional framework in the region, see S. Vinogradov, "Transboundary Water Resources in the former Soviet Union: Between Conflict and Cooperation," 36 *Natural Resources Journal,* 393: 411.

11. TACIS is the European Union's program for Technical Assistance to the Commonwealth of Independent States (Armenia, Azerbaijan, Belarus, Georgia, Kazakhstan, Moldova, Russia, Tajikistan, Turkmenistan, Ukraine, and Uzbekistan).

12. The project was negotiated during spring 1998. The five States have formally requested GEF assistance for the Aral Sea Basin program in April 1996. The aforementioned project with a total cost of $21.3 and a GEF contribution of $12 million is likely to constitute the single most important operation at the regional level for the years 1998–2003. Cofinancing figures are $9.3 million that is distributed as follows: $4.1 million (five Central Asian States), $2.8 million (Government of Netherlands), $1.4 million (European Union/TACIS), and $1.0 million (Swedish International Development Agency).

13. See NRMP's website, at http:// www.nrmp.uz/water_and_environment.htm (last visited, September 2002).

14. The European Union assists the States of the Former Soviet Union (FSU) through the Technical Assistance to the Commonwealth of Independent States (TACIS).

15. While it had been unclear whether these draft agreements would remain as three separate agreements or be merged into a single instrument, it has been decided to opt for the former solution.

16. Text of the convention reprinted in 36 *I.L.M.*700 (1997). None of the Central Asian countries have signed yet to become a party to the UN Convention.

17. The States have made it clear that they do not want the agreements to be considered as outputs of the project, as the time frame for their adoption may be longer than the duration of the project.

18. About 12.5 percent of the Aral Sea Basin Program's water resources originate in the country; only a fraction is used for irrigation. It contributes between 3 and 5 km^3 water per year to the Amu Darya.

19. The treaties concluded between Afghanistan and the former USSR—to which the five Central Asian republics are successors—provide inter alia for regular exchange of technical information and for the adoption of joint measures to prevent changes in the course of frontier waters. On these treaties, see M. Nanni, "The Aral Sea Basin: Legal and Institutional Issues," 5 *Review of the European Community and International Environmental Law,* 130, 131.

20. See IWMI website at http://www.cgiar.org/iwmi/casia/casia.htm (last visited, September 2002).

21. There are over 80 water reservoirs, 45 hydropower plants, and 57 large dams in the Aral Sea Basin.

22. Water Related Vision for the Aral Sea Basin for the Year 2025, UNESCO/ SABAS, France, 2000, p. 50.

23. A practice that was also in force during the pre-independence period, see Nanni, *supra,* 131-132.

24. Information drawn from the USAID website at http://www.usaid.gov (last visited, November 2002).

25. The GEF Operational Strategy indicates, inter alia, that

The overall strategic thrust of GEF-funded international waters activities is to meet the agreed incremental costs of (a) assisting groups of countries to better understand the environmental concerns of their international waters and work collaboratively to address them; (b) building the capacity of existing institutions (or, if appropriate, developing the capacity through new institutional arrangements) to utilize a more comprehensive approach for addressing transboundary water related environmental concerns; and (c) implementing measures that address the priority transboundary environmental concerns. The goal is to assist countries to utilize the full range of technical, economic, financial, regulatory, and institutional measures needed to operationalize sustainable development strategies for international waters.

26. MP Wat/2000/6/Add 1 (Economic Commission for Europe/UN/ECE).
27. MP Wat/2000/6/Add 1 (Economic Commission for Europe/UN/ECE).

BIBLIOGRAPHY

Documents and Reports of International Organizations

European Commission (TACIS) (1997). *WARMAP Project: Formulation & Analysis of Regional Strategies on Land and Water Resources.*

UNESCO/SABAS (2000). *Water Related Vision for the Aral Sea Basin for the Year 2025.* Paris: United Nations Educational, Scientific, and Cultural Organization.

United Nations Environmental Programme (2000). *Regional Environmental Report on Aral Sea.* New York: UNEP.

Articles

Bedford, D. P (1996). "International Water Management in the Aral Sea Basin." *Water International* 21: 63–69.

Glantz, M. H. (1998). "Creeping Environmental Problems in the Aral Sea Basin." In: I. Kobori & M. H. Glantz (eds.) *Central Eurasian Water Crisis: Caspian, Aral and Dead Seas.* Tokyo: United Nations University Press: 25–52.

Kirmany, S. & G. L. Moigne (1997). "Fostering Riparian Cooperation in International River Basins—The World Bank at its Best in Development Diplomacy." *World Bank Technical Paper.* 335.

Micklin, P. (2000). *Managing Water in Central Asia.* London: Royal Institute of International Affairs. CACP Paper.

Micklin, P. (1998). "Regional and International and Responses to the Aral Crisis: An Overview of Efforts and Accomplishments." *Post-Soviet Geography and Economics.* 39 (7): 399-417.

Nanni, M. (1996). "The Aral Sea Basin: Legal and Institutional Issues." *Review of the European Community and International Environmental Law*. 5 (2): 130–37.

O'Hara, S. L. (2000). "Lessons from the Past: Water Management in Central Asia." *Water Policy* 2: 365-384.

Vinogradov, S. (1996). "Transboundary Water Resources in the Former Soviet Union: Between Conflict and Cooperation." *Natural Resources Journal*. 36: 393–414.

Appendix

TABLE 10. Aral Sea Basin Program Donor Contributions (as of 09/25/02)

Amounts in $million

Name of the Program/Project	Source	Contributions Grants	Loans	Comments
Program 1				
Regional Water Resources Management Strategy	NTF GEF EU	2.2 4.4 6.5		Fundamental provisions of water management strategy in the Aral Sea Basin developed.4 legal Agreements drafted.
Dam Safety and Reservoir Management	SWE USAID	1.2 1.6		10 dams surveyed; early warning system and monitoring equipment procured. Syr Darya cascade Agreement signed.
Subtotal Program 1		13.9		
Program 2				
Hydrometeorological Services (Transboundary Monitoring Stations)	SWISS UK	1.8 0.2		Swiss Government additional commitments as contributions were not included.
Regional Environmental Information System	EU NOR EU	1.7 0.1 6.5		Regional database established.
Subtotal Program 2		10.3		

(continued)

Table 10. Aral Sea Basin Program Donor Contributions (as of 09/25/02) *continued*

Amounts in $million

Name of the Program/Project	Source	Grants	Loans	Comments
Program 3				
Water Quality Management	NTF	1.0		
Uzbekistan Drainage	PHRD	1.0		Environment assessment
	NTF	0.3		and detail design of the first phase completed.
Subtotal Program 3		2.3		
Program 4				
Wetland Restoration	NTF	3.8		Pilot project at Lake
	NATO	0.24		Sudoche implemented.
Restoration of Northern Part of the Aral Sea	ITA	0.5		Design completed.
	WB		83.3	
	IFAS	0.25		
Environmental Studies in the Aral Sea Basin	NTF	0.7		Studies carried out.
Subtotal Program 4		5.5	83.3	
Program 5				
Water Supply, Sanitation, and Health—Uzbekistan	NTF	0.3		Development of water supply is successful.
	WB		75.0	
	SWISS	5.5		
	KfW		9.4	
	KFAED		19.8	
	DN	0.3		
Water Supply, Sanitation, and Health—Turkmenistan	WB		30.3	Development of water supply is successful.
Water Supply, Sanitation, and Health—Kazakhstan	KFAED		11.5	Development of water supply is successful.
	KfW		7.7	
	WB		7.7	
Water Supply, Sanitation, and Health—Tajikistan	SWISS	0.3		Project launched.
	WB		30.0	
Subtotal Program 5		6.4	161.1	

(continued)

TABLE 10. Aral Sea Basin Program Donor Contributions (as of 09/25/02) *continued*

Amounts in $million

Name of the Program/Project	Source	Grants	Loans	Comments
Program 6				
Integrated Land/Water	NTF	1.0		Studies carried out.
Management in the	Finland	0.3		Pilot projects proposed.
Upper Watersheds	Turkey	0.36		
Subtotal Program 6		1.7		
Program 7				
Operational Water	CIDA	1.5		Four head works
Resources Management	FR	0.2	4.0	equipped.
	US	0.2		
	IFAS	0.11		
Subtotal Program 7		2.0	4.0	
Program 8				
Capacity Building of the	NTF	0.5		EC-IFAS operational.
Interstate Institutions	UNDP	1.5		Water Management
	CIDA	1.6		Training Center
				established.
Subtotal Program 8		3.6		
Total		**45.7**	**278.7**	

CIDA	Canadian Development Agency	NTF	Netherlands Trust Fund
DN	Government of Denmark	PHRD	Japanese Policy and Human
EU	European Union		Resources Development Fund
FR	Government of France	SWE	Government of Sweden
GEF	Global Environment Facility	SWISS	Government of Switzerland
ITA	Italian Trust Fund	UK	Government of United Kingdom
KFAED	Kuwait Fund for Arab Economic Development	US	Government of United States (USAID)

Chapter 6

Conclusion: Globalization, Multi-Governance, and Transboundary River Basin Management

Matthias Finger, Ludivine Tamiotti, and Jeremy Allouche

The main object of this conclusion is to discuss and develop the concept of multi-level governance with regards to transboundary river basin management. As shown in the introduction, the nation-state is no longer the most appropriate unit for managing unilaterally transboundary river basins. Indeed, in all of the different case studies, one can already observe a multitude of actors involved in river basin management (e.g., nongovernmental actors). These new actors bring with them new perspectives and interests. Besides, one can also witness dynamics at different levels, from the global to the local. More precisely, it appears that the economic, political, or ecological issues linked to water resources management cannot be analyzed and solved without taking into account the current dynamics at the four major levels, i.e., the global, the regional, the national, and the local levels. In this regard, the four case studies have clearly shown the necessity for a multi-level approach to governance, be it in water or elsewhere.

We consider governance to be a function for collectively solving problems. As stressed in the introduction, water problems in particular, cannot be adequately understood simply by looking at one level (e.g., the national level). Therefore, the management and solution of any given problem (e.g., water) cannot be limited to just one level, and a multi-level approach is required. Moreover, the originality of our approach lies in the fact that governance is seen as a dynamic process by which the different levels (from local to global) continuously interact along the policy circle, i.e., from problem definition to evaluation.

Applied to transboundary river basin management, our approach aims at understanding the different ways in which water resources are currently being managed. In doing so, one needs to look at the various actors' interests and the different perspectives of how the river should be managed at each level. Furthermore, all river basins are characterized by a number of issues ranging from political, via economic to ecological ones. Our approach identifies the different actors and their perspectives at each level and then ties them to the various functions necessary for managing water resources.

This conclusion will be divided into two parts. First, each case study will be summarized and analyzed in order to outline the different problems encountered, namely in terms of actors' interest, conflicting issues, and contradicting dynamics between the various levels. Second, we will link the four case studies to our concept of multi-governance as developed in the introduction and by doing so will provide a conceptualization of water multi-governance. This conceptualization will be done by identifying and applying the different functions and phases of the policy circle to transboundary rivers. In this respect, our approach to governance is quite different from the current theories of governance, introducing an additional dynamic view.

1. SUMMARY AND ANALYSIS OF THE DIFFERENT CASE STUDIES

In this part, the four case studies will be summarized and their main characteristics recalled. At the end, the main issues and actors will be presented graphically. Of course, some simplifications will be necessary in order to draw a general picture for each case study, as there are a large number of issues interacting at the same time. General issues such as political pressures, economic interests, and biophysical dimension of the management of a river basin will be considered. The most crucial issue at stake in the management of a given river basin will be identified and examined. Finally, we will compare the case studies and identify different categories of cases, in terms of actors, issues, and levels.

1.1. Summary of Each Case Study

1.1.1. The Mekong River Basin

The management of the Mekong River Basin is widely determined by its biophysical characteristics. Capacity to utilize waters for hydropower changes according to geographic zones. Concerns at stake are inter alia the following: sustainable development, forest management, protection of biodiversity, quality of water, and quantity of the river's flow. Its management is then mainly determined by two issues: protection of water quality and of its ecosystem, as well as the regulation of the water flow. The first issue can be explained by the extremely rich biodiversity of the basin, while the second one pertains to the fact that the basin is a major hydropower source.

The historical context is relevant for the understanding of the current problems because of a strong influence of communism on local populations. This partly explains why these local populations are usually not very much involved in, and concerned with, the management of the river basin. The war history of this region must also be considered in order to explain the present degradation of the ecological, social, and economic situation. The basin's rich cultural heritage and diversity is also an essential characteristic of this river basin that cannot be neglected.

Most riparian countries are developing policies that intend to improve the navigation of the Mekong River and to expand the number of projects with regards to hydroenergy use. However, some of them also adopt approaches often oriented toward sustainable development and the management of the consequences of uncontrolled industrial development on the environment and on biodiversity.

The Mekong River, in terms of levels involved, can be characterized as follows: first, the *regional level* is playing a specifically relevant and interesting role in the definition of the needs of the riparian States in terms of water allowances, water quality, and biodiversity protection. The activities of the Mekong River Commission (MRC) are essential in policy planning for sustainable development, utilization, management, and conservation of water and related resources of the Mekong River Basin. Regrettably, the commission does not have any influence on the final outcomes of the projects or even on the stages beyond the early investigative phase projects. Moreover, one major actor, i.e., China, is not a member of the commission and policy planning for the entire basin at the regional level cannot really be done without China's participation.

Despite the fact that the regional level is now playing a more preponderant role in water planning, the *national* level still remains important in that it is active along two priorities: on the one hand ecological sustainability and on the other hand the promotion of major economic development projects. Despite the creation of the MRC, there are still some competing claims between the different governments over the development of irrigated land and hydropower development.

The process of financing dams construction or navigability improvements projects is taken on by international organizations, foreign
aid institutions, and loan agencies, i.e., by the *global level*. By determining the amount of funding, the global level also determines the exact
definition of the projects. It seems that international institutions favor
the development of hydropower and transportation projects because
such projects are easier to finance and implement.

The *local level* also plays a fundamental role in the management
of the Mekong River Basin. Indeed, nongovernmental organizations,
mainly international NGOs, are used for the implementation of projects
financed by international organizations, mainly because of their skills
and field knowledge. However, locally based NGOs are very limited in
number and in influence because of the lack of established public participation processes and infrastructures.

Figure 14 summarizes the issues and stakes of the Mekong River Basin.

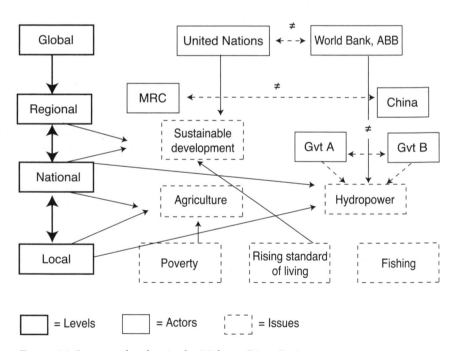

Figure 14. Issues and stakes in the Mekong River Basin

1.1.2. The Danube River Basin

The Danube is the second longest river in Europe, and it flows through seventeen different countries. This makes the Danube a very atypical case. In addition to that, there are a large number of constructions (mainly hydroelectric plants and reservoirs) all along the Danube that makes the management of this river basin even more complex. The Danube River also supports the supply of drinking water, agriculture, industry, fishing, tourism, power generation, navigation, and the end disposal of wastewater of the riparians. The particularly rich biodiversity explains a pronounced regional and global interest regarding the governance of this area.

The historical context is also especially relevant, since most of the countries concerned experienced the influence of communism. This might have had consequences, among others, on the political involvement of local populations. Furthermore, the influence of the social context is a significant factor in the governance of the Danube River. There is indeed substantial participation of the local populations in the management of the river basin and in the monitoring process of the quality of waters. The main issues at stake in the Danube are therefore the problem of water quality and more broadly the consequences of the intensive agricultural, industrial, and urban uses of the river.

This river basin does have a local, national, and regional dynamics. The *regional level* is mainly concerned with the cooperation between States through information and cooperation mechanisms provided by international institutions, such as the Danube Commission or the Environmental Action Program for Central and Eastern Europe. The *national level* is more focused on the definition of domestic and foreign policies. It deals, for instance, with the building of dams, with the protection of the quality of waters and of the ecosystem, and with the problems of minorities. There is some strong connection between this level and the regional level, mainly because of the European Union access issue. Indeed, for some countries the perspective to become accepted as a member of the European Union exerts a strong influence on the definition and implementation of policies for the protection of the environment. Finally, the *local level* also plays an important role in the management of the Danube River Basin. Indeed, many nongovernmental organizations are involved in the management of this river basin, either by promoting public and government awareness on environmental issues or by assisting in the development of appropriate policies.

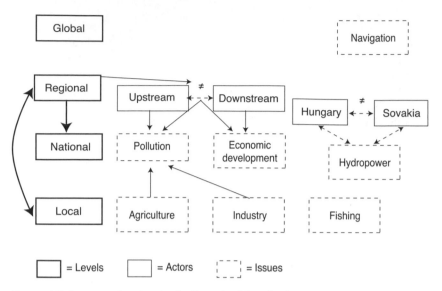

FIGURE 15. Issues and stakes in the Danube River Basin

Figure 15 summarizes the issues and stakes of the Danube River Basin.

1.1.3. The Euphrates River Basin

The biophysical dimension is relevant in the analysis of the major problems of management encountered in the Euphrates River Basin because its topographic and climatic characteristics are rather different from one riparian to another, putting upstream countries in a predominant position. The Euphrates being mainly used for irrigation purposes, the main issue at stake is primarily one of fair allocation of water among riparians. Its management is being complicated by the fact that seasonal distribution of waters generally does not coincide with crop needs. Moreover, some parts of the river are characterized by a rapid flow, whereas in some other areas the river's flow weakens. All of this creates important inequalities in the allocation of water among the three concerned countries.

Though the biophysical characteristics of this river basin are essential, the historical and geopolitical dimensions are much more fundamental. The history of the management of the Euphrates River is characterized by various degrees of "hydropolitical tensions" depending on the political context (e.g., the former cold war environment or the Kurdish question); economic needs (e.g., the respective policies of the three riparians toward the amount to be drawn from the Euphrates); and geopolitical stress (e.g., the difficult relationships between leaders of Turkey and Syria). The historical context is essential in order to understand the present difficulties in the Euphrates River Basin. Indeed, the influence of the Soviet Union, especially during the cold war, and the political consequences of the Gulf War are two determinant factors of the present political tensions between the riparian countries. However, the geopolitical context is the most important element. The relationships between the three riparians are based on coexistence. There is no actual cooperation and no process of exchange of information. This context creates situations where water allocation issues lead to crisis situations, especially between Syria and Turkey. Instead, the influence of the social context is rather limited, since the local populations do not have access to democratic processes or structures. However, it must be noticed that the local populations are quite knowledgeable as to how to deal with the irregular flows of the river and how to deal with periods of scarce water resources.

The decisions regarding the management of this river basin are therefore entirely driven by the *national level*. No other institutions have a specific role in the management of the Euphrates River Basin. The major national interests are the development of hydroelectric power, agriculture, and industry. The three relevant ministries—Agriculture, Water, and Foreign Affairs—are basically in charge of these interests. Three factors are determining the shape of these interests, i.e., economic concerns, foreign policy, and domestic policy. Projects developed by the riparians over the river are dams to produce electricity, constructions aiming at preventing flooding, or dams for irrigation purposes. The *regional level* does not have any relevance since there is no cooperation between the riparians, except on some issues of security (e.g., the Kurdish issue) and trade. There are no regional institutions to which the three riparians commonly belong and each country has adopted a different position on the 1997 Convention on Non-Navigational Uses of International Waters. Furthermore, the *global level* is also not relevant, since the financing of water management projects is only assumed by the national level. Indeed, institutions such as the World Bank would refuse to finance any projects where there are such tensions between riparians.

Figure 16 summarizes the issues and stakes of the Euphrates River Basin.

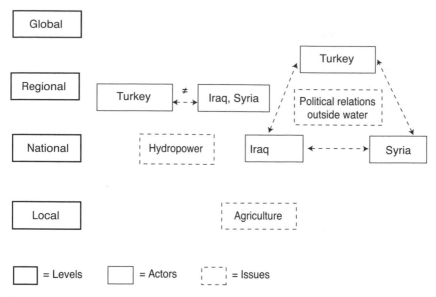

FIGURE 16. Issues and stakes in the Euphrates River Basin

1.1.4. The Aral Sea Basin

The main biophysical characteristics are as follows: the Aral Sea is the world's fourth-largest inland body of water and its protection has been of great concern mainly because of its remarkable shrinkage in volume during the second half of the twentieth century. This has usually been described as one of the biggest ecological catastrophes of the past century. The management of this sea is entirely dependent upon two international rivers, namely the Amu Darya and the Syr Darya, which are the Aral Sea's main sources of inflow. The Aral Sea Basin is composed of the five former Soviet republics (Kazakhstan, Kirghizstan, Tajikistan, Turkmenistan, and Uzbekistan) and Afghanistan. Rather than the depletion of the sea itself, it is the management of the two international rivers, which has, over the years, become the focus of attention.

The historical and political context is indeed determinant, since the Aral Sea basin has experienced the centralized management of Soviet planners, who decided in the 1960s to divert large quantities of water from the Amu Darya and the Syr Darya for irrigation purposes. This has had obvious consequences on the lack of political involvement of the local populations. The present geopolitical context is also highly relevant, since the riparians have only emerged recently as independent countries with different views with respect to water use (e.g., between upstream and downstream countries). Given the former Soviet management, the influence of the social context is rather nonexistent for the Aral Sea Basin.

The national, regional, and global levels are therefore the major dynamics within this basin. Given the biophysical dimension, the *regional level* is particularly essential. The regional level is characterized by an accumulation of mechanisms of cooperation for the management of the Aral Sea. The *national level* should not be neglected, since the countries connected with the Aral Sea are newly independent States, which are rather reluctant to commit themselves to stringent international agreements. Instead, the *global level* is very much present in the management of the basin. Indeed, international agencies, such as UNDP, UNEP, the World Bank, and the European Union are greatly involved in the financing of projects related to the Aral Sea. The global level furthermore has a key role, because it encourages the development of both the local and the regional levels. International agencies indeed tend to favor the development of regional institutions for the management of the Aral Sea, while they also condition their assistance on the promotion of public awareness, and participation.

Figure 17 summarizes the issues and stakes of the Aral Sea Basin.

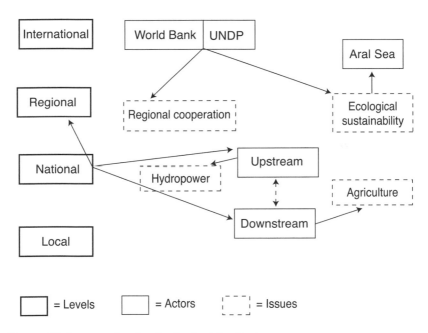

Figure 17. Issues and stakes in the Aral Sea Basin

1.2. *An Analysis of the Four Case Studies: Issues, Actors, and Levels*

Although each of these four case studies is unique (especially regarding historical and geopolitical aspects), one can still compare the different issues, actors, and levels involved.

In terms of *issues*, one can distinguish the case studies dominated by major political stakes from the ones dominated by economic/ecological ones. The Euphrates and the Aral Sea are clearly characterized by major *political stakes*. As outlined by Arnon Medzini and Aaron T. Wolf, there are four main factors that are determining the relationships between the three riparian States: the respective policies of the three States regarding the amount of water to be drawn from the Euphrates and Tigris rivers, the Kurdish question, the rivalry between the Iraqi and the Syrian branches of the Ba'ath Party, and Syria and Turkey's historical animosity (and especially Syria's claims to Turkish territories).[1] Water management is here surely an issue of security. The Euphrates is almost a closed system, within which the three riparians are trying to obtain their share of water. Yet, even if political pressure is high, economic considerations are progressively becoming more important. As summarized by Arnon Medzini and Aaron Wolf, "Since there is no cooperation between the riparians for the common development of the river basin, the individual economies can and will solve their problems independent of their riparian neighbors in the global system which, in future, will lead to closer economic and political co-operation between the States."[2]

The same can be said for the Aral Sea Basin, whose stakes are also currently driven by politics. Indeed, the collapse of the former Soviet Union has led the five new independent republics to put water at the top of their political agenda, leading to a shift in the relationships among them. The independence of the republics forced them to be concerned with transboundary water management issues. However, independence also meant for each republic a different and not necessarily compatible way to manage their water resources. We are now in a situation where each government is setting different priorities ranging from reducing their dependence on energy supply for upstream States to developing irrigation schemes for downstream States. Political stakes do not exclusively drive the management of this river basin however. Other physical, environmental, economic, and social factors must also be taken into account. In fact, the Aral Sea Basin represents a major economic stake in that cotton production is a significant source of revenue for these countries. Yet, the economic development around the Aral Sea Basin, be it for hydroelectric or irrigation purposes, cannot fully take place unless this region reaches a certain political stability. Therefore, even though water has a clear economic importance, water remains primarily a source of political conflict.

Economic/ecological issues are clearly the most important issues for the two other case studies, i.e., the Mekong and the Danube rivers. In the Mekong case, national authorities no longer consider the river basin as a political issue but rather as a tool of economic development and ecological sustainability. In part due to the influence of international organizations, the focus in the different projects is now on the promotion of sustainable development. Since the physical characteristics of this river basin have significant potential as a major hydropower energy source,[3] cooperation over this issue is highly attractive to the six riparian States. Simultaneously, these countries are now also increasingly focusing on the conservation of the biodiversity (genetic, flora, and fauna) and on the various terrestrial, wetland, and aquatic ecosystems.

As for the Danube, this is probably the case where the political tensions around the management of the river are the lowest of the four case studies examined, and this despite the legal dispute between Hungary and Slovakia over the Gabcíkovo-Nagymaros dam project. In fact, the seventeen riparian countries have a more down-to-the-earth view of the management of this river, focused as it is primarily on economic interests. A large number of dams, dikes, navigation locks, and other hydraulic structures have been built for navigation or hydroelectric power purposes and this, all along the river and especially in Germany and Austria.[4] Because of such economic use, the basin is also facing serious problems of quality, quantity, and reduced biodiversity problems, thus threatening the health of the overall basin's ecosystem.[5] As outlined by Stephen McCaffrey, the current management of the river can be characterized by two trends, which will increase the protection of the Danube and its ecosystem: enhanced cooperation between the different actors and the strengthening of the various institutions dealing with the river basin.[6] However, the long history of overexploitation and pollution of the Danube, due to intensive agricultural, industrial, and urban uses, still has to be addressed.

In terms of *actors*, one can interestingly notice that the two cases, which are dominated by major political stakes (i.e., the Euphrates and the Aral Sea) are also characterized by the predominance of a single type of actor, namely national government. In the Euphrates River Basin, the situation is most simple in that the only major actors involved are the three respective governments of Iraq, Syria, and Turkey. In fact, this can partly be explained by geopolitical reasons and by difficult biophysical conditions. Indeed, the amount of water available in the

Euphrates River varies considerably from month to month and from year to year.[7] Even if this river basin only involves three riparian States—in comparison, for example, with the seventeen countries covered by the Danube River Basin[8]—they are regularly experiencing hydropolitical tensions. In the case of the Aral Sea Basin, one can identify two major actors involved, namely international organizations and national governments. However, the different national governments still control water policy firmly, since water constitutes, to a certain degree, a guarantee of their national independence and a way of affirming their sovereignty. In this context, the role of international organizations is of course key, as it is international organizations that force governments to cooperate. There are of course a certain number of international institutions involved but their role is clearly limited for various reasons. First, the national governments are quite reluctant to give away their sovereignty over water resources and these institutions therefore do not receive sufficient financial and technical support by the respective governments in order to carry out their objectives. Second, all of these institutions do not have a clear mandate in order to make them a major player in the management of this basin.

In the Mekong and Danube cases, one can see on the other hand a situation where there are many actors involved. At all levels, an important number of actors, ranging from international organizations, river basin commission, national and local governments, and nongovernmental organizations are found. All of these actors have of course divergent interests but they have managed to develop a certain cooperation. In the Mekong case, most of the involved countries still remain strongly influenced by the former communist political system, yet are all in a process of economic reform, increasingly combined with a more sustainable approach to the river basin development. However, China's interest in the downstream water resources allocation has been so far very limited, even though downstream riparians might suffer from dam constructions in China. This does not facilitate a cooperative and sustainable development of the river basin, as conflicts on equal sharing and water benefit may arise. At this point, political tensions may actually come back.

In terms of *levels*, two different approaches to transboundary river basin management are distinguished, i.e., a top-down and a bottom-up approach. Interestingly, all of the cases in developing countries are characterized by a top-down approach. The role of the international organization is certainly a plausible explanation for such top-down approaches given that they have a considerable influence on policy formulation and financing.

The Danube case is in fact the only bottom-up case. Of course, the national and regional levels are also an important feature in the governance of this river basin but both are actually pushing for a more grassroots approach. The riparians of the Danube are clearly following the current trend in water resources management, which emphasizes that "water development and management should be based on a participatory approach, involving users, planners and policy-makers at all levels" (Principle two of the Dublin statement).[9] The Danube Environment Programme is a very good illustration of a supranational initiative pushing for the development of a bottom-up approach. In fact, this programme favors the emergence of a grassroots approach by supporting institutional development, capacity building, and NGO involvement. In the Danube case, one can also see a trend that delegates new responsibilities to the local public authorities, rather than to the national ones. Local municipalities clearly play an important role in managing the river basin, be it in terms of defining or in terms of implementing regulatory programs. Moreover, and unlike the three other cases studies examined, many initiatives in the Danube River Basin case involve civil society. As recalled by McCaffrey, "Every person in the Danube River Basin has several possible roles—as a consumer of goods and services; as a producer of waste at home [or] at the workplace; as a user of recreation facilities; and as a citizen whose choices and actions express cultural, social, aesthetic, spiritual and environmental values. NGOs are established by members of the public to promote public and governmental awareness of environmental issues and to assist in the development of appropriate policies."[10] Moreover, civil society is not just an important actor in itself, but can also largely influence the decision-making process. In this regard, McCaffrey cites the example of the Gabcíkovo-Nagymaros case as a good illustration of the potential significance of action by civil society: "Hungarian citizen protests against the project being constructed by Hungary and Czechoslovakia on the Danube were an important factor contributing to Hungary's eventual decision to withdraw from the project. The protests were focused on the project's effects on the environment."[11] In sum, "It seems likely that institutions of civil society will continue to develop in the Lower Basin countries and will have an increasing impact upon governmental decision making in relation to the sustainable development of the Danube Basin."[12] The Danube is therefore the only major case in this book that is clearly moving toward a much more decentralized approach to river basin management, going far beyond traditional nation-state-oriented policies and management.

The three other cases are clearly different in that the national and international levels are more predominant. In the Euphrates, for example, the local level is simply not involved. As stressed by Medzini and Wolf,

> While generally it is an oversimplification to speak of states as homogeneous entities, it is actually fairly accurate in the case of the Euphrates riparians. Both Syria and Iraq have autocratic, authoritarian regimes and, while the Turkish government is a representative parliamentary system, the major water authorities ... are fairly autonomous and do not rotate with each government. Moreover there are no regional institutions to which the three belong—Turkey is a member of NATO, while Syria and Iraq belong to the Arab League; each has a different position on the 1997 Convention on Non-Navigational Uses of International Waters; and, with the exception of Turkey, there is little in the way of non-governmental political activity."[13]

In the Aral sea case, international organization and bilateral donors are encouraging a more grassroots participatory approach involving civil society. But, as stressed by Laurence Boisson de Chazournes, "Public participation at all levels of water management should also be encouraged, although it should be recognized that for the region, it is a new issue to be dealt with and that today, such a process remains in a nascent phase."[14] The Mekong case could be included in the bottom-up approach. Indeed, as stressed by Natana Gajaseni, Oliver William Heal and Gareth Edwards-Jones, "People are beginning to assert their right to participate in their own governance and they have become important actors in water resource management."[15] However, "Overall, the local organisations and the NGO sector, together with intergovernmental organisations at the national level, are not yet significantly developed to contribute effectively to comprehensive governance in resource management in the basin."[16]

1.3. Conclusion

The economic value of water is greatly increasing since the middle of the twentieth century, especially due to the development of hydro-electricity and irrigation. These developments have led, on the one hand to an increasing competition over water between the different users and, on the other hand, to an increasing pressure on water resources. Both of these developments have had important consequences on the governance of the four case studies examined in this book. These developments have been perceived as an increasing threat to the good social and economic development of the region and have actually led to two opposite considerations: one consideration is to regard water as an issue of national security, where the national government is the sole and only entity responsible for managing water resources; the other consideration is to see that increased interdependence leads to increased cooperation. But in any case, the transnational character of the river examined significantly complicates their management. Different countries are involved and each national (or local) government's decisions directly affects the other riparian countries. Therefore, the governments involved are forced to interact. Such interaction can be characterized by a cooperative attitude (as in the Mekong River case except for China) or a noncooperative attitude (as in the Euphrates cases). Such *interdependency* among governments has still increased because of the diversification of water uses and because of the growing consumption of water. In terms of governance, the necessity for resolving possible conflicts over water use therefore appears critical.

The second major conclusion one can draw from the different case studies examined is that river basin management is becoming much more *complex*, mainly because the management of the basin is influenced by new actors such as international organizations and NGOs. These actors, as highlighted in the introduction, are reshaping the international system at its four major levels (international, regional, national, and local) by calling into question the traditional state-centric way public affairs have been managed so far, as they are more generally putting into question the nation-state as the sole and only actor. This reconfiguration can also be witnessed in most of our case studies where there is now a multitude of actors influencing the way in which international rivers are managed. In terms of governance, these new actors are claiming the right to also participate in the formulation and implementation of policies. It seems that the multiplication of actors is an important factor in facilitating a more cooperative attitude over river basin management. The role of international organizations in this process of course should not be underestimated. In turn, this more cooperative attitude among the different actors also leads to a much more integrated approach to the basin, involving economic development, ecological sustainability, and political stability. In the case of the Danube and the Mekong rivers, the creation of river basin commission certainly also constitutes an important factor for the realization of such an integrated approach.

Finally, our four case studies have clearly highlighted the fact that there is simultaneously a *policy* and *institutional dimension* to governance. Both of these dimensions have of course already been put forward in the previously discussed governance theories. For instance, the important role of river basin commissions has already been highlighted in regime theory. On the other hand, common property resources management scholars argue that more sustainable policies result from clear rules about water rights and allocations. For both theories, good governance is seen to be instrumental for achieving sustainable river basin management. However, both theoretical approaches are still essentially static in terms of institutional design and deterministic in terms of policy formulation. They do not adequately capture the current situation in the river basins and underestimate, in our view, the changing paradigm, as well as the institutional dynamics, in river basin management. This is why we wish, based on our four cases, to make a contribution to theory-building in (multi-level) governance.

2. Conceptualization of Governance

Governance basically aims at solving collective problems by means of involving the relevant stakeholders, for example, in the case of river basin management. Traditionally, governance therefore involves both actors (e.g., the relevant stakeholders) and policy (e.g., the policy cycle), as represented in Figure 18.

However, this is in our view a too limited approach to governance, given that the different policies are not really distinguished. Some would of course argue that levels are already implied by the different actors. But levels should not simply be seen as a subcategory of actors or policies. Levels clearly have a dynamics of their own. The local level, for example, is rather influenced by tradition. It is also the level most recognized for its legitimacy and for its effectiveness since it is at the local level where most problems occur. The national level has in fact a dual function: to provide (national) solidarity and identity; and a function of external representation given that it is the national government that is recognized as the main legal entity at the international level. The

Figure 18. Traditional representation of governance

regional level has generally been neglected but is now increasingly acknowledged as being important. The nature of its dynamic is much debated. Is it simply an (interstate) response to globalization or a product of globalization? In fact, this level is very much built on the idea of economic integration and efficiency gains. The global level is also very difficult to characterize. In some ways, it may be seen as the sum of the three previous levels described. Yet, it has also a dynamic and an existence of its own.

In our case studies, all of these levels were present. In the cases of river basin management, the *local* level is mainly faced with ecological problems, whether in terms of quality or quantity. In both cases, the local level is still very much counting on the national level for support. In the Danube case, it is increasingly the regional level that takes on the role of supporting the local level, for example, through the river basin commission. When it comes to the *national* level, one can see that governments play a key role in water resources management, given in particular that water clearly is viewed as an issue of national security. From there, it follows either a cooperative or a conflictual attitude with respect to the other riparians. The *regional* level, furthermore, is particularly important in water resources management, given that it is seen, by many, as the ideal level where water resources can be managed in a sustainable way. Of course, this level is very much influenced by the different governments, yet is slowly becoming independent from governments. This is the case of the Danube, where the river basin commission has established close relationships with the local level without the mediation of the national level. Finally, the *international* level has a clear influence on river basin management whether in terms of financing or of policy formulation.

Accordingly, both from a conceptual and from an empirical point of view, one can conceptualize multi-level governance in Figure 19.

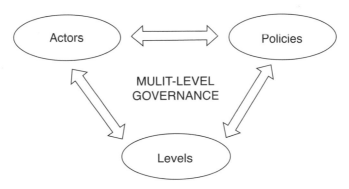

FIGURE 19. Multi-level governance

In fact, our approach to governance enables us to have a much more dynamic view of collective problem solving of river basins. With this multi-level approach, governance is not a static phenomenon but can be seen as a sequence of steps in the policy cycle. As a next step, we therefore need to better understand and define what the different steps of the policy circles are.

Despite their differences, all four case studies share some common features. All of these features pertain to the various policy functions that need to be performed if governance is to be successful. In particular, policy issues need to be identified, and policies need to be formulated, monitored, implemented, and evaluated. Often, all of these functions are not explicitly attributed to one or several actors, but generally all them are performed in one way or another, sometimes by a combination of actors.

If one now applies this model to our four case studies and in particular introduces the four different levels (global, regional, national, and local), the complexity of transboundary river basin governance becomes rapidly apparent. Figure 21 tries to summarize this complexity.

Figure 21 clearly shows the never-stopping dynamic within the governance of a river. Priorities over the *policy issues* are periodically called into question by the various actors involved, even though such questioning does generally not follow a logical policy circle (e.g., implementation, monitoring, and evaluation). As a matter-of-fact, most of the actors—from international organizations to local NGOs—to different degrees, have the opportunity to question the policy or to reformulate the issues, sometimes even without having adequately monitored or evaluated the outcomes of the policy.

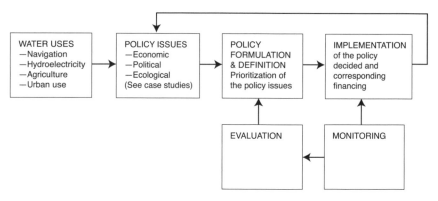

Figure 20. Major policy functions

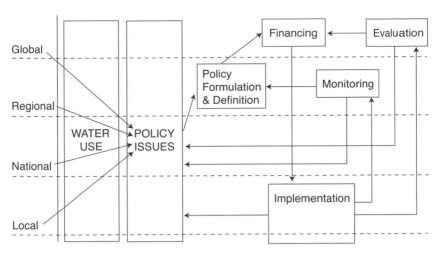

FIGURE 21. Multi-level governance of transboundary rivers

Policy formulation and definition is the response to conflicting demands and interests over water use. Interestingly, one can see that in the cases with primarily economic/ecological stakes (i.e., the Danube and the Mekong), the initiative from policy definition stems from the regional level. This is mainly the role of the river basin commissions. The Danube is even more complex, as the local level is also involved in such policy definition through a participatory approach. In the cases where water is still an object of political conflict, however, policy definition remains at the national level, given that it is primarily considered to be a national security issue. This is particularly true for the Euphrates case. In the Aral Sea, the national level also remains the decisive level for policy definition, although it is strongly influenced by major international financial institutions such as the World Bank. If one looks at the different policy issues, it appears that only at a global level are these issues identified and approached in the most comprehensive way. This is also the level where policy formulation is most inclusive, although not always most pertinent, given that at times local participation is being left out.

In most cases, *financing* also comes from the international level. Three of the four case studies are located in developing or transition countries and therefore heavily rely upon the help of the international community. This further exacerbates the fact that international funding agencies often play a powerful role in the entire policy cycle, i.e., from policy formulation to evaluation. For example, in the case of the Mekong, the international funding agencies were clearly pushing for economic development (hydropower), and this despite the increasing pressure for ecological sustainability.

In all of the cases, *implementation* is always at the local level. It is usually in this implementation phase that the relationships between the different riparians can shift from a 'cooperative' to a more conflictual approach. We here use the term *conflict* in a broad sense, now going from diplomatic unrest to military mobilization. If the policy was formulated at the regional level, the probability of conflict is however rather low. On the other hand, the potential for conflict is highest if the implementation relies exclusively on the nation-state (e.g., the Euphrates case), be among nations or between the national government and its local citizens.

Monitoring instead generally occurs at the regional level, especially when it is at this level that policies where initially formulated and defined. On the other hand, *evaluation* is clearly an international function, given the fact that it is the funding agencies that are generally interested in evaluating the proposed policy.

In short, by combining policy with actors and levels, we have been able to develop a complex model capable of accounting for the dynamics that are currently taking place in the four river basins observed. Our conceptual model of multi-level governance goes further, we think, than the currently still prevailing approaches of regimes or common property resources management.

3. CONCLUSION

The governance of transboundary rivers is becoming increasingly complex. This book has highlighted the fact that the nation-state can no longer be the single actor responsible for transboundary river basin management, as governance of transboundary rivers can no longer occur at one single level. Of course, an approach that goes beyond the different aspirations and policies of the nation-states and tries to involve the major stakeholders at all levels in order to define and implement a more efficient and ecologically more sustainable approach to transboundary water governance, is still far from being achieved. In this regard, the concept of multi-level governance can be used as an important tool to highlight the different dynamics within and between each level, as well as to identify the main functions that need to be performed in transboundary water governance. Moreover, our approach clearly introduces a dynamic perspective, and this in contradiction to the currently predominant theories of governance.

Although the concept of multi-level governance has been developed from, and applied to, selected river basins, multi-level governance is a much broader concept and approach, and as such is not limited to integrated river basin management. To recall, we have developed the idea of multi-level governance against the background of globalization and the declining role of the State in managing public affairs. In this regard, transboundary river basin management is just one illustration of the growing need for, and pertinence of, a new multi-level governance approach in water and in other transboundary natural resources issues more generally.

NOTES

1. Medzini, A. & Aaron T. Wolf, "The Euphrates River Watershed: Integration, Coordination, or Separation?" p. 110.

2. Ibid., p. 133.

3. N. Gajaseni, O. W. Heal, & G. Edwards-Jones, "The Mekong River Basin: Comprehensive Water Governance," p. 45.

4. McCaffrey, S. "The Danube River Basin," p. 80.

5. Ibid., p. 81.

6. Ibid., p. 97.

7. Medzini & Wolf, "The Euphrates River Watershed," p. 106.

8. McCaffrey, "The Danube River Basin," p. 79.

9. International conference on water and the environment (1992), *International conference on water and the environment: Development issues for the 21st century*, http://www.wmo.ch/web/homs/icwedecc.html.

10. McCaffrey, "The Danube River Basin," p. 90.

11. Ibid., p. 90.

12. Ibid.

13. Medzini & Wolf, "The Euphrates River Watershed," p. 118.

14. Chazournes, L. B. "The Aral Sea Basin: Legal and Institutional Aspects of Governance," p. 164.

15. N. Gajaseni, O. W. Heal & G. Edwards-Jones, "The Mekong River Basin," p. 58–59.

16. Ibid., p. 61.

Contributors

JEREMY ALLOUCHE is currently Director of the Water Institutions and Management Competence Centre at the Swiss Federal Institute of Technology (EPFL). He holds a Ph.D. from the Graduate Institute of International Relations in Geneva. Allouche has published several monographs and articles on water management and conflict including: *Water Privatisation: Transnational Coporations and the Re-regulation of the Water Industry* and "La géopolitique de l'eau en Asie centrale: De la colonisation russe à la conférence d'aide à l'Afghanistan."

LAURENCE BOISSON DE CHAZOURNES is Director and Professor of the Department of Public International Law and International Organization at the Faculty of Law, University of Geneva. She has been a visiting scholar at the University of Michigan and Georgetown University and a visiting professor at the Graduate Institute of International Studies in Geneva. Besides teaching and research, her professional experiences include consulting activities with the World Bank where she was a principal counselor on international and environmental law. She has also worked with specialized UN agencies in this field. She has published a large number of articles and books on international organizations, and international, environmental, and water law.

GARETH EDWARDS-JONES is Professor of Agriculture and Land Use Studies at the School of Agricultural and Forest Science at the University of Wales. His research interests range from the ecology and conservation of species on farm land and economics of invasive species, to economics of marine-protected areas and has published a large number of scientific articles and books on these subjects.

MATTHIAS FINGER is Professor at the Swiss Federal Institute of Technology (EPFL). Before coming to EPFL, he was Professor of Management of Public Enterprises at the Swiss Graduate Institute of Public Administration (IDHEAP), an Associate Professor at Columbia University, and Assistant Professor at the Maxwell School of Citizenship and Public Affairs at Syracuse University. He holds a Ph.D. in Political Science and a Ph.D. in Adult Education (both from Geneva). He is the author or coauthor of twelve books and over eighty book chapters and scientific articles, mainly in the area of social and organizational change. Besides teaching and research, he also is a consultant with the Swiss Postal Service, the Rockefeller Foundation, and various UN agencies. His main interests pertain to organizational and institutional change and innovation, as well as to questions of governance.

NANTANA GAJASENI is Assistant Professor in the Biology Department at Chulalongkorn University. Her research interest focuses on freshwater ecology, natural resource management, and ecology. She is currently working on several funded projects on the Mao Klong River and has worked on projects funded by the European Community and by USAID and has published several books and scientific articles.

OLIVER WILLIAM HEAL is Honorary Professor at the University of Edinburgh and Visiting Professor at the University of Durham. He is also former Director of the Institute of Terrestrial Ecology. His research interests focus on ecology and ecosystems and has published a large number of articles and books on the subject.

STEPHEN McCAFFREY is Distinguished Professor and Scholar at the McGeorge School of Law, University of the Pacific. He is former chair of the International Law Commission (ILC). As the ILC's "special rapporteur" for international watercourses, he guided the ILC's work that formed the basis of the 1997 United Nations Convention on the Law of the Non-Navigational uses of International Watercourses. McCaffrey currently serves as legal consultant to the Nile River Basin Cooperative Framework, a UN-sponsored project whose aim is to forge a multinational agreement on utilization of the Nile's water resources. He has argued in front of the International Court of Justice, taught environmental law courses in German at Swiss universities, advised the State Department, represented foreign governments in river-use disputes, and has published several books and more than fifty articles in law journals.

ARNON MEDZINI works at the Department of Geography at the Oranim School of Education in Israel. His research interests are the geography of Israel and the Middle East, water, and the Bedouins rivers, and has published a large number of articles on the Euphrates and Jordan.

LUDIVINE TAMIOTTI is Legal Affairs Officer in the Trade and Environment Division of the World Trade Organization (WTO) in Geneva. She holds advanced law degrees from the Universities of Aix-en-Provence, Geneva, and New York. She held posts at the United Nations International Court of Justice in The Hague, the Institute of Advanced Studies in Public Administration in Lausanne, and the Environment Unit of the United Nations High Commissioner for Refugees in Geneva. In the WTO, she conducts research and provides legal advice on TBT and trade and environment issues, as requested by members. She has also been Co-Secretary to the panel on EC—Trade Description of Sardines. She is involved in technical assistance activities on both technical barriers to trade and trade and the environment.

AARON T. WOLF is Associate Professor of Geography in the Department of Geosciences at Oregon State University. He has an M.S. in water resources management and a Ph.D. in land resources, (the latter emphasizes policy analysis) from the University of Wisconsin, Madison. His research focuses on issues relating to transboundary water resources and to political conflict and cooperation, where he combines environmental science with dispute resolution theory and practice. He has acted as consultant to the U.S. Department of State, the U.S. Agency for International Development, and the World Bank on various aspects of international water resources and dispute resolution. He has been involved in developing the strategies for resolving water aspects of the Arab-Israeli conflict, including coauthoring a State Department reference text, and participating in both official and "track II" meetings between coriparians. He is (co-) author or (co-) editor of six books, and close to fifty journal articles, book chapters, and professional reports on various aspects of transboundary waters. Wolf currently coordinates the Transboundary Freshwater Dispute Database.

Index

France, 110, 134
French model, 22

Gabcikovo-Nagymaros, 84–86, 87,
    183, 185
GAP project, 109, 114, 116, 117,
    118, 120, 129, 130
Gas, 113, 116, 149, 160
Germany, 79, 80, 81
Gleick, Peter, 27
Globalization, 1–4, 6, 14, 19
Global Water Partnership, 22
Global Environmental Facility, 56,
    155, 162
Gorbachev, Mikhail, 147
Government, 1, 8
Governance, 1–3, 43, 95, 97, 173,
    185, 187, 188–193
    Comprehensive, 55, 58, 59, 61,
        71, 72–74, 186
    Definition, 1, 6
    Multi-level governance, 4–12,
        32, 173, 188–193
    Global governance, 5–6
    Good governance, 7–8, 188
    Local governance, 10–11
    National governance, 9–10
    Participatory governance, 91,
        97, 185, 191
    Regional governance, 8–9
Green Revolution, 26
Gulf War, 109, 115, 116, 122, 123,
    179

Harmon doctrine, 30, 134
Hatay, 110, 111, 114
Helsinki Convention on the Protec-
    tion and Use of Transboundary
    Watercourses and International
    Lakes, 83, 91
Himalayas, 45
Hungary, 84, 89, 90, 96, 185
Hydroenergy/electricity/power (see
    also dams), 24, 27, 33, 45, 51,
    53, 54–55, 58, 64–65, 67, 70, 72,

80, 84, 86–87, 119, 123, 126,
    128, 148, 159, 174, 175, 179,
    183, 186, 192
Hydrology, 108

India, 25, 134
Indus Water Treaty, 134, 135
Industry, 119, 123, 179
Italy, 79
Integrated water/river basin manage-
    ment, 22, 23, 33, 56, 57, 58, 64,
    71, 87, 89, 93, 94–95, 160, 164,
    187
Interdependency, 3, 12, 64, 86, 187
International Boundary and Water
    Commission, 16
International Commission for the
    Protection of the Rhine against
    Pollution, 16
International Commission for the
    Protection of the Danube River,
    83, 97, 98
International Court of Justice, 84,
    85
International Fund for the Aral Sea,
    152, 153, 164
International Joint Commission, 16
International Law Association, 21
International Monetary Fund, 7,
    62
International organizations, 3, 22,
    62–63, 74, 81, 88, 164, 176, 183,
    184, 186, 187, 190
International Water Management
    Institute, 158
Interstate Commission for Water
    Coordination, 152, 153, 155, 157,
    164
Iraq, 30, 103–146
Iran, 25, 115, 117, 118, 123
Iran-Iraq War, 32, 117, 123
Irrigation (see also agriculture), 16,
    17, 19, 26, 27, 70, 89, 104, 111,
    126, 129, 147, 159, 161, 162,
    175, 178, 180, 182, 186

SUNY series in Global Politics
James N. Rosenau, Editor

American Patriotism in a Global Society—Betty Jean Craige

The Political Discourse of Anarchy: A Disciplinary History of International Relations—Brian C. Schmidt

Power and Ideas: North-South Politics of Intellectual Property and Antitrust—Susan K. Sell

From Pirates to Drug Lords: The Post–Cold War Caribbean Security Environment—Michael C. Desch, Jorge I. Dominguez, and Andres Serbin (eds.)

Collective Conflict Management and Changing World Politics—Joseph Lepgold and Thomas G. Weiss (eds.)

Zones of Peace in the Third World: South America and West Africa in Comparative Perspective—Arie M. Kacowicz

Private Authority and International Affairs—A. Claire Cutler, Virginia Haufler, and Tony Porter (eds.)

Harmonizing Europe: Nation-States within the Common Market—Francesco G. Duina

Economic Interdependence in Ukrainian-Russian Relations—Paul J. D'Anieri

Leapfrogging Development? The Political Economy of Telecommunications Restructuring—J. P. Singh

States, Firms, and Power: Successful Sanctions in United States Foreign Policy—George E. Shambaugh

Approaches to Global Governance Theory—Martin Hewson and Timothy J. Sinclair (eds.)

After Authority: War, Peace, and Global Politics in the Twenty-First Century—Ronnie D. Lipschutz

Pondering Postinternationalism: A Paradigm for the Twenty-First Century?—Heidi H. Hobbs (ed.)

Debating the Global Financial Architecture—Leslie Elliot Armijo

Political Space: Frontiers of Change and Governance in a Globalizing World—Yale Ferguson and R. J. Barry Jones (eds.)

Crisis Theory and World Order: Heideggerian Reflections—Norman K. Swazo

Political Identity and Social Change: The Remaking of the South African Social Order—Jamie Frueh

Social Construction and the Logic of Money: Financial Predominance and International Economic Leadership—J. Samuel Barkin

What Moves Man: The Realist Theory of International Relations and Its Judgment of Human Nature — Annette Freyberg-Inan

Democratizing Global Politics: Discourse Norms, International Regimes, and Political Community—Rodger A. Payne and Nayef H. Samhat

Landmines and Human Security: International Politics and War's Hidden Legacy—Richard A. Matthew, Bryan McDonald, and Kenneth R. Rutherford (eds.)

Collective Preventative Diplomacy: A Study of International Management—Barry H. Steiner

International Relations Under Risk: Framing State Choice—Jeffrey D. Berejikian

Globalization and the Environment: Greening Global Political Economy—Gabriela Kütting

Sovereignty, Democracy, and Global Civil Society—Elisabeth Jay Friedman, Kathryn Hochstetler, and Ann Marie Clark

Imperialism and Nationalism in the Discipline of International Relations—David Long and Brian C. Schmidt (eds.)

United We Stand? Divide and Conquer Politics and the Logic of International Hostility—Aaron Belkin

Globalization, Security, and the Nation State: Paradigms in Transition—Ersel Aydinli and James N. Rosenau (eds.)

Mediating Globalization: Domestic Institutions and Industrial Policies in the United States and Britain—Andrew P. Cortell